COMMUNICATING

YOUR RESEARCH

with

MEDIA

Sara Miller McCune founded SAGE Publishing in 1965 to support the dissemination of usable knowledge and educate a global community. SAGE publishes more than 1000 journals and over 800 new books each year, spanning a wide range of subject areas. Our growing selection of library products includes archives, data, case studies and video. SAGE remains majority owned by our founder and after her lifetime will become owned by a charitable trust that secures the company's continued independence.

Los Angeles | London | New Delhi | Singapore | Washington DC | Melbourne

COMMUNICATING

YOUR RESEARCH

with

SOCIAL

MEDIA

A Practical Guide to Using Blogs, Podcasts, Data Visualisations and Video

Amy Mollett
Cheryl Brumley
Chris Gilson
Sierra Williams

⑤SAGE

Los Angeles | London | New Delhi
Singapore | Washington DC | Melbourne

Los Angeles | London | New Delhi
Singapore | Washington DC | Melbourne

SAGE Publications Ltd
1 Oliver's Yard
55 City Road
London EC1Y 1SP

SAGE Publications Inc.
2455 Teller Road
Thousand Oaks, California 91320

SAGE Publications India Pvt Ltd
B 1/I 1 Mohan Cooperative Industrial Area
Mathura Road
New Delhi 110 044

SAGE Publications Asia-Pacific Pte Ltd
3 Church Street
#10-04 Samsung Hub
Singapore 049483

Editor: Jai Seaman
Assistant editor: Alysha Owen
Production editor: Victoria Nicholas
Copyeditor: Sarah Bury
Proofreader: Sharon Cawood
Marketing manager: Sally Ransom
Cover design: Shaun Mercier
Typeset by: C&M Digitals (P) Ltd, Chennai, India
Printed and bound by
CPI Group (UK) Ltd, Croydon, CR0 4YY

© Amy Mollett, Cheryl Brumley, Chris Gilson, and Sierra Williams 2017

First published 2017

Library of Congress Control Number: 2016953042

British Library Cataloguing in Publication data

A catalogue record for this book is available from the British Library

ISBN 978-1-41296-221-6
ISBN 978-1-41296-222-3 (pbk)

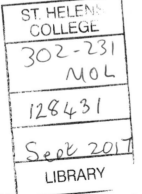
At SAGE we take sustainability seriously. Most of our products are printed in the UK using FSC papers and boards. When we print overseas we ensure sustainable papers are used as measured by the PREPS grading system. We undertake an annual audit to monitor our sustainability.

CONTENTS

AUTHOR BIOGRAPHIES

The four authors contributed equally to this volume.

Cheryl Brumley is Senior Producer at *The Economist* where she produces and reports on economics, politics and science for their podcasts on *Economist Radio*. Previous to this job, Cheryl worked for the LSE Public Policy Group, producing the award-winning podcast series the *LSE Review of Books*, as well as podcasts for the *LSE Impact Blog*, the *British Politics and Policy Blog* (2010) and *EUROPP – European Politics and Policy* (2012). Additionally she has worked at the BBC World Service. Cheryl is also a freelance radio journalist reporting for outlets such as Public Radio International and Deutsche Welle English. Cheryl was named a 'New Voices' scholar for her achievements as a producer by the Association of Independents in Radio (AIR). With her co-authors, Cheryl has won a Times Higher Education Award for Knowledge Exchange. She tweets @cherylbrumley.

Chris Gilson is Managing Editor of *USAPP – American Politics and Policy*, the blog of the LSE's United States Centre (http://blogs.lse.ac.uk/usappblog/). He also launched and managed the *LSE's British Politics and Policy Blog* (2010) and *EUROPP – European Politics and Policy* (2012), and supports the creation and management of other blogs around the LSE. He has a Master's degree in Geography and a postgraduate diploma in Strategic Management, all from the University of Waikato, Hamilton, New Zealand. With his co-authors, Chris has won a Times Higher Education Award for Knowledge Exchange. His interests include blogging, research communication, US politics, urban politics and community activism. He tweets @chrishjgilson.

Amy Mollett is Social Media Manager at the London School of Economics. She previously managed several blogs at LSE, including *LSE Review of Books* and the *LSE Impact Blog*. She has published popular guides for researchers on using social media, including 'Using Twitter in university research, teaching and impact activities: A guide for academics and researchers' with Patrick Dunleavy, downloaded by 100,000 readers. With her co-authors, Amy has won a Times Higher Education Award for Knowledge Exchange. Amy is a graduate of the London School of Economics and the University of Sussex, and is interested in digital engagement, social media trends and academic communication. She tweets @amybmollett.

Sierra Williams is Community Manager at PeerJ, an open access science publisher for biology, medicine and computer science. She previously managed strategic communications for the global non-profit, Open Knowledge International and spent four years at the London School of Economics as the Managing Editor of the LSE Impact Blog, one of the most popular LSE blogs, now reaching over 90,000 readers a month. She has an MPhil in Conflict Resolution and Reconciliation Studies from Trinity College, Dublin, and a BA in Sociology from the University of San Francisco. Sierra also provided research and administrative support for the LSE Impact of Social Sciences Project. Her interests are in open scholarship, the role of expertise in society, distributed communication networks and data sharing. She tweets @sn_will.

ACKNOWLEDGEMENTS

We would like to thank the following people:

LSE PUBLIC POLICY GROUP AND LSE BLOGS

Patrick Dunleavy, Jane Tinkler, Simon Bastow, Nazreen Fazal, Anthony McDonnell, Natalie Allen, Sonali Campion, Joel Suss, Ellie Harries, Mark Carrigan, Julian Kirchherr, Avery Hancock, Helena Vieira, Rose Deller, Kieran Booluck, Stuart Brown, Tena Prelec, Sean Kippin, Luke Temple, Artemis Photiadou, Ros Taylor, Roch Dunin-Wasowicz, Richard Berry, Carl Cullinane, Paul Rainford, and Danielle Moran.

THE LSE BLOGGING COMMUNITY

Thank you to the many people who have supported the blog projects at LSE over the years: the Research Division, Library Services, Arthur Wadsworth and Ben Durant in IMT/Web Services, Jane Secker, Matt Lingard and Chris Fryer in LTI, the Communications Division, and all the academics and researchers who have come to make blogging and using social media a part of their communities.

PEOPLE WHO HAVE GIVEN CONTRIBUTIONS TO THE BOOK

Mikko Jarvenpaa, CEO of Infogram; Valerie Chepp and Lester Andrist at The Sociological Cinema; Zach Wise, creator of TimelineJS; Dr Rachel Aldred and Martin Key from British Cycling; Gordon Katic, *Cited* podcast; Claire Navarro, *Hold That Thought* podcast; Amiera Sawas, participatory photo project; Mike Yorke, The Ho: the People of the Rice Pot; Mark Blyth, Professor of Political Economy at Brown University; and Wagner Oliveira and all at FGV DAPP.

THEORETICAL INFLUENCES AND ACADEMIC HEROES

Many thanks to all the external contributors to the LSE Blogs and in particular the broad range of research and opinion shared in the LSE Impact Blog, which has helped us enormously in thinking through and reflecting on our own working practice (many of whom are cited in these pages!). In particular, we'd like to acknowledge the work of Martin Weller, Matt Lingard, Melonie Fullick, Jane Secker, Cassidy Sugimoto, Bonnie Stewart, Andy Tattersall, James Wilsdon, Andy Miah, Cameron Neylon, Melissa Terras, Stephen Curry, and Pat Thomson.

SAGE

Many thanks to Katie Metzler, Jai Seaman and Alysha Owen for guiding this book to publication.

FAMILY AND FRIENDS: CHRIS

A book like this has many contributors beyond the authors, and it would be remiss of me not to mention those without whom this project would not have been possible. My amazing wife Sally, for providing support far above and beyond the call of duty every day. My newly arrived son Max, who was busy gestating at the same time as much of this book. My parents, Dorothy and Clive, for inspiring me to write (and write well). Jane Tinkler and Patrick Dunleavy, who took a punt on academic blogging (and me!) all those years ago, and helped to make it shine and become an inspiration to many. And my fantastic co-authors: Amy for keeping us on the straight and narrow throughout (and to deadlines), Cheryl for her ninja-like editing skills, and Sierra for her deep knowledge and insight of theories of research that helped make this book so much more than a simple 'how to' manual.

FAMILY AND FRIENDS: SIERRA

I'd like to thank my own personal research lifecycle: Many thanks to social media for its daily inspiration and frustration in my life. A big thank you to co-authors, Amy, Cheryl and Chris for being all-around awesome humans and for all their effort in this collaboration (GoogleDocs FTW!). I'd like to thank Jane Tinkler, Patrick Dunleavy and Simon Bastow for organising the primary research and community upon which our work was built. And also thanks to all LSE Impact Blog contributors who have over the last four years helped me work through what research dissemination and communication is and could be. To my Mom, Dad and Brooke for their thoughtful and caring engagement with me over the years and their supportive outreach. And finally, to Sam for his positive impact in my life and for making all those cups of tea.

FAMILY AND FRIENDS: CHERYL

First and foremost, my husband, Marcus Yorke, who, in addition to providing me with emotional support and encouragement throughout the book-writing process, made sure I was sufficiently fed with his scrumptious dinners. To my family: my sister Michelle; my dad, Mike; mom, Kathy; Aunt Eleanor; James; Sarah; Rich: and Henry, who live in our native swampland, Florida, across the pond, but whose love and support transcend that distance. To the Yorkes: Jessie, Mike and Valerie, for their endless encouragement. To my friends: Brigitte, Sally, Amiera and Gee for inspiring me. To Emily Kasriel, for showing me the radio ropes at the Beeb. And finally to my co-authors, Amy, Chris and Sierra: you three are my academic spirit animals and the best co-authors and friends anyone could ask for.

FAMILY AND FRIENDS: AMY

My eternal gratitude to my family for all their support while writing, and in life generally: thank you to Gina for sharing great ideas from the art world that have made it into this book; to Casey for helping me find confidence on the bad days; to Dad for putting the belief in his girls that they can do anything; to Mum for nurturing my interest in the social sciences; and to Martin, who will be forever reminding us that we handed in the manuscript on the day Leicester won the Premier League. Thank you to my friends for their wild encouragement, especially Asiya, Becca, Chloe, Francine, Frieda, Julia, Maitrayee and Nick. Thank you to Ben for his love, support and generosity. A special thanks to Jane Tinkler and Patrick Dunleavy for being such inspiring mentors, and to my co-authors, Cheryl, Sierra and Chris, who blow my mind on a daily basis with their ideas, passion for what they do, and friendship.

THE COMPANION WEBSITE

Designed to help students bring their research to life with social media, *Communicating Your Research with Social Media* is supported by a variety of online resources at https://study.sagepub.com/Mollett to help you maximise your research's impact and highlight its relevance.

- **Author videos** provide top tips on how to use the advice from this book in your own research and offer further insight into how you can combine social media with academic practice.
- **Blog post guides** demonstrate how social media have changed the way real-world research can be communicated to reach wider, more engaged audiences.
- **Links to published podcasting and twitter guides** give you additional support in using these social media in research and other academic settings.
- **A Twitter hashtag** gives you an opportunity to connect directly with the authors of the book and its other readers so you can ask questions, keep the conversation going and create a network of researchers.

INTRODUCTION

WHY DID WE WRITE THIS BOOK?

This book is your guide to how to use social media to share what you do and what you learn with the world. As part of a team working in higher education academic communication and public engagement in the early 2010s, it became clear to us that there was a growing appetite and interest in using online and multimedia methods to share academic work, interact and engage with non-academic users. Academics, students, communications professionals and others were coming to us to seek advice on how to use social media to share and promote their work. Much of this impetus has come out of the rise of the impact agenda – as exemplified in the UK by the Research Excellence Framework (REF) – and indeed, calls for researchers and academics from funding bodies, government and the media to be more engaged with the public have never been louder and clearer. These more external calls come alongside wider academic-led conversations that reflect dissatisfaction with the traditional mechanisms for scholarly communication.

In light of this, we saw the need for an accessible, practical handbook that any academic, researcher or communicator working in public engagement of research could use to start out on their own and plan their projects – we could have used one when we began our own work in this area! This book brings together much of what we've learned over the past five years. We have enjoyed creating successful projects at the London School of Economics, and we want to put our knowledge and experiences into one place so that it can be useful for academics and researchers who are faced with similar creative challenges.

Anyone looking at the higher education, think tank, or any other information-led sector in much of the world can see that the research process has changed and is still changing. It is now becoming apparent that social media, in the form of blogging, podcasting, data visualisations and social networks will have an increasing part to play in how we share research and how the public and specific audiences want to engage with it. We have also seen polarisation between the social sciences and the humanities, and the Science, Technology, Engineering and Mathematics (STEM) fields, with much of the public's focus and resources on the latter. We think that both the social sciences and STEM fields are working towards the same goals – to maximise the worth and impact of their work on society – and as such the techniques and tools we discuss in this book will be useful to knowledge workers across all of these fields.

WHO IS THE BOOK FOR?

This book is for people we call 'knowledge workers'. This includes academics, researchers, students and communications professionals looking to understand more about research. It also covers anyone at knowledge-led institutions, such as think tanks, NGOs, charities and nonprofits. You also don't have to be tied to an institution for this book to be useful; if you're an independent scholar, you'll get a great deal out of it as well.

While we discuss in detail the academic context, this is not itself an 'academic' book. We look at the conceptual, theoretical and historical influences that have shaped the social media environment, but we do so from a largely practice-based approach that can be easily applied and remixed in a variety of settings. The sources we draw on range from journal articles to blog posts to journalistic commentary to industry perspectives to radio interviews, all aiming to give the reader a familiar and pragmatic overview of how various social and digital media tools operate and function in the world.

HOW TO USE THIS BOOK AND WHAT WE HOPE PEOPLE WILL GET OUT OF IT

There is no one way to use this book, and no need to read it in chapter order. You may want to skip ahead to a content-focused chapter that you're most interested in if you're working on a specific project, or you might be interested in a general overview of social media before jumping into more detail. Most of all, we want you to use this book for ideas and inspiration. As such, we've made a point of providing useful hints and tips to help you to start using social media and social networks to promote your work and the research that you do.

While we are only too aware that social media is an incredibly fast-moving landscape (this book would have been very different if we'd written it in 2005, 2010 or even 2014), we want this to be as much of a timeless guide as possible. With that in mind, we'll talk about the history of social media and the various platforms, give examples of the impacts they can and have had, and how to make and maximise your own. While platforms will come and go, and new social media trends will arise, this guide is written so that it can be used with any future platforms. We want this book to help you to develop your knowledge, skills and understanding of academic communication, public engagement and working in research in an increasingly digital world.

In Chapter 1 we look at what social media are, why they matter, and what they can do. Beginning with a history of social media, from the formative years of the ARPANET and email, we look at how social media have evolved to become the important force for communication and collaboration for knowledge workers today. We take a close look at social media's influence across society, from the Arab Spring and #blacklivesmatter to human rights, business and health. We then take an in-depth look at the role of social media in education and research, focusing on its use by think tanks, research bodies, academics and researchers. We close out the chapter with some critical reflections on the role of social media and academic engagement.

In Chapter 2 we outline the shape of research today and propose an inclusive framework for understanding research communication: the Research Lifecycle. We first look at the theoretical underpinnings of digital scholarship and how this fits alongside wider conceptual understandings of research and society. Following a brief exploration of various models for understanding communication and research, we present a Research Lifecycle Framework, which breaks the research process down into six distinct, yet often overlapping, phases:

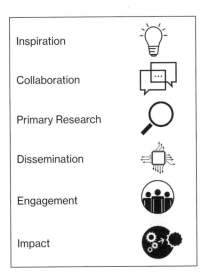

Inspiration

Collaboration

Primary Research

Dissemination

Engagement

Impact

We explore how these six phases currently operate in practice and how social media are already being used in support of these activities. Given the limitations of existing models for understanding modern research practice, this framework aims to capture the complexity of the research environment and provides a way to explore how certain media can act in the interests of both researchers and wider society, signposting examples and media formats to be discussed in greater detail in the following chapters. With examples and ideas for academics, researchers and communications professionals working in all fields, and signposted with the symbols above, our Research Lifecycle Framework will guide you through the changing landscape of how research is done and operates in the world.

In Chapter 3, the first of our practical chapters, we look at how to create and maintain a successful blog. After first looking at what makes blogs unique and different from websites, we cover the intellectual ancestors of blogs from American Revolutionary pamphlets to the British suffragette movement of the 19th and 20th centuries. We then turn to why blogging is important for academics and knowledge workers, and the kind of impacts it can have. Following this, we present an in-depth discussion of how to create a successful blog and blog posts, including a look at writing styles, different blog formats and platforms.

In Chapter 4 we consider how infographics and data visualisations can be an exciting and effective way to share your research with new audiences. From the data visualisations created by Florence Nightingale in 1859 showing soldier mortality rates in the Crimea hospital camps, to the viral infographics created by cycling campaigners that have shaped government policy in the UK today, we break down the biggest moments in the history of data visualisation and explore some of the most impactful examples of recent times. We then guide you through making decisions about which infographics and data visualisations to use for your project, how to make them using free online tools, and when to hire professional designers.

In Chapter 5 we focus on podcasts and look at the myriad ways you can use the medium of audio to carry out or disseminate research. We first charter the podcast journey from a Harvard research centre in the early 2000s to the *Serial* podcast craze in 2014 before exploring its history as an academic tool. We then look at creative uses of podcasting in 'Academic Soundscapes', using examples from our experiences of podcasting at LSE, of producing original content on research ranging from the sociology of London's Chinatown to a podcast on ground-breaking research on the London riots of 2012. Chapter 5 also includes a full range of examples from podcasters in the knowledge-producing sphere: from a professor who produced podcasts on immunology while on sabbatical in Africa, to a pair of researchers at Yale who record interviews on pedagogy with other staff at Yale University. These rich examples, along with a comprehensive overview of podcast formats and advice on how to create and develop a successful series, seek to give the novice researcher the tools and inspiration needed to jump-start a podcast series of their own.

In Chapter 6 we look at creating photos and videos to use on social media, blogs and websites. We have all witnessed the rise of photo- and video-driven social media platforms and apps like Instagram, Snapchat and Pinterest, as well as the success of Facebook's own native video function. But what does this trend mean for academics and researchers who are looking to share work with new audiences? This chapter looks at the history of photos and videos in academic and research-based projects and traces linkages through to examples of innovative photo and video use in research and dissemination projects with social media today. We then cover the growing body of literature which is already showcasing how photo- and video-driven social media platforms are becoming effective communication tools for research projects, before looking at some step-by-step guides for your own work.

Bringing the book to a close, in Chapter 7 we look at how to pull together a social media and digital engagement strategy. Given the stretched nature of researcher workloads and research budgets, getting the most out of your social and digital media activity is incredibly important. We look at how individuals and research teams can understand their own communication efforts, situate goals and objectives, and act strategically. We look at (1) how research can approach existing platforms, (2) the many different ways of measuring various social activities, and we end with (3) a brief reflection on how to navigate the pitfalls and risks of social media. This chapter is useful for a project management view of integrating social media across your research project, but also offers insights for those interested in crafting an individual online identity and what it means to take part in a networked public sphere. We also investigate the more negative aspects associated with social media use, including trolling, harassment and abuse, and how to challenge the echo chamber effect of social media.

WHO WE ARE AND OUR WORK

By now, you might be asking questions about who we are, and what our background is that informs the writing of this book. With that mind, here's some context about us and the work that has been done at the London School of Economics around social media. We all met at the LSE's Public Policy Group (a small research think tank led by Professor Patrick Dunleavy and Jane Tinkler), where we initially worked on the PPG's various academic blog platforms. While our work began with blogs, it quickly expanded into the various social media platforms we discuss in this book. What started with a need to promote the content of our blogs with Twitter and Facebook grew and continued with Pinterest posts, podcasts and infographics and data visualisations.

BLOGGING AND SOCIAL MEDIA AT THE LONDON SCHOOL OF ECONOMICS

Our story began in 2010, with the launch of what became the LSE's British Politics and Policy blog by the LSE's Public Policy Group. The blog began as *Election Experts* and was aimed at providing academic commentary and coverage of what was to be an important General Election. This blog was a reaction to the growing plethora of political blogs, such as the now-infamous *Order Order* blog run by Guido Fawkes (the pseudonym of blogger, Paul Staines) and Will Straw's *Left Foot Forward*, and more mainstream blogs from 'old media' such as *The Guardian* and *The Evening Standard*.

With many commentators predicting a hung parliament at that point, PPG felt that the new blog could act as a 'referee' for the various policy positions being set out by the major parties. Employing a mostly full-time editor, Chris Gilson, co-author of this book, the blog would leverage academic expertise across the LSE to raise the level of debate about the coming election by publishing evidence-based commentary. After the election, the blog continued to develop and grow its readership, reaching about 200,000 visitors in its first year of operation, and soon became the highest ranked, and the second most read, university blog in the UK. This blog effort was to become the basis for eight similar blogs established at the LSE over the next five years.

As of 2016, the LSE has developed blogs aimed at disseminating academic commentary on politics and policy issues:

- *British Politics and Policy* (2010). Mentioned above.
  http://blogs.lse.ac.uk/politicsandpolicy
- *EUROPP: European Politics and Policy* (2012). Initially a vehicle for European academics and experts to explain the eurocrisis plainly to a wide audience, EUROPP publishes academic commentary on contemporary politics and policy for the European Union (EU) and the wider European region.
  http://blogs.lse.ac.uk/europpblog
- *USAPP: American Politics and Policy* (2012). Academic commentary giving the views of both insiders and outsiders on US politics, policy and elections.
  http://blogs.lse.ac.uk/usappblog
- *LSE Brexitvote* (2015). Academic commentary and 'refereeing' of the debate over the UK's membership of the European Union in relation to the 2016 'Brexit' referendum.
  http://blogs.lse.ac.uk/brexit

The LSE has also developed blogs which are focused on knowledge exchange and academic impact:

- *The Impact of Social Sciences Blog* (2011). Posts daily insights and commentary pieces from academics focused on best practice for achieving impact with their research. As of November 2015, the Impact Blog has over 90,000 visitors every month and is regarded as one of the most important resources for academics in this area.
  http://blogs.lse.ac.uk/impactofsocialsciences

- *LSE Review of Books* (2012). Radically reduces the speed at which social science books are reviewed and makes reviews accessible outside paywalls so that they are free to access worldwide. Uses a wide pool of reviewers, including PhD students, early career researchers, and senior academics and emeritus professors.
  http://blogs.lse.ac.uk/lsereviewofbooks
- *LSE Business Review* (2015). An effort to provide a university-led base for knowledge exchange between business and academics.
  http://blogs.lse.ac.uk/businessreview

In September 2011, the LSE Public Policy Group produced *Using Twitter in university research, teaching, and impact activities* (Mollett et al., 2011), a guide for academics and researchers who were interested in learning more about using Twitter, including setting up an account, building up followers and using Twitter in research projects. It has been downloaded by over 100,000 researchers all over the world.

In June 2012, the LSE Public Policy Group was the winner of a Times Higher Education Management and Leadership Award, in the category of Knowledge Exchange/Transfer, for its blogs. The awards highlight best practice by university professional services staff in areas which include student services, human resources, library services and strategic planning. One of the panel judges stated of the LSE's blogs:

> Their high-quality blog highlights real and important issues. ... The comments and debates it generates influences stakeholders and policymakers in a much more subtle and powerful way than traditional lobbying. This is a real example of how social scientists do have, and can demonstrate, real impact.

PPG PODCASTS

Alongside its blog offerings, PPG also produced podcasts from 2012 to 2014. By 2012 the LSE was already well known for its public event podcasts, with a large number of well-known speakers, which were played millions of times by listeners every year. These event podcasts are recordings of LSE-hosted events. They are limited to the topic of a single lecture and the ensuing question and answer sessions. By producing podcasts in-house, PPG was able to expand the LSE's podcast offerings, displaying a wider range of podcast formats, mostly radio magazine-style, across four of its blogs.

Our four series of podcasts took advantage of the recent growth in podcast listenership, which has been exemplified by offerings such as *This American Life* and *Radiolab*. Averaging about 25–30 minutes in length, our podcasts were also much shorter than the LSE's event podcasts, the aim being for them to be of a manageable length to cover a listener's commute or other short activity. In just a few years, the *LSE Review of Books* garnered a lot of attention in the podcasting community, including in 2012 when it was voted the top UK academic podcast in the European Podcast Awards.

The podcasts worked in concert with their 'home' blogs, with blog authors as hosts, and with each podcast having a home page on their blog where they could be presented with links to references and further reading materials.

Now that we've talked about what you can expect from this book and where we are coming from, it's time to move on to a question that's sure to be on your mind by now: why do social media matter, and what do they offer for my work and my research?

1

SOCIAL MEDIA: WHY THEY MATTER AND WHAT THEY CAN DO

There are now a billion social-media posts every two days ... which represent the largest increase in the capacity of the human race to express itself at any time in the history of the world.

– King, 2014

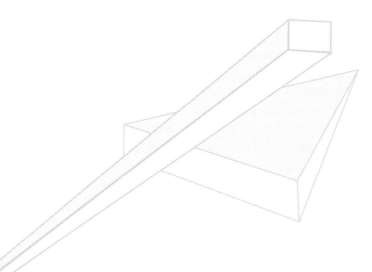

Monday 7am. Kira wakes up with her alarm. It was a long weekend of mostly writing – she's trying to get the revisions to a paper she's co-authoring back to the journal's editors this week. Checking her emails on her iPhone while she waits for her coffee to brew, Kira's excited to see an email pop up from the editor of a journal she submitted to late last year; her article is now available online! It won't be in the print edition for another six or maybe even nine months (there's a huge backlog, the editor says), but she now has the link for her piece. As she heads out the door, Kira posts a link to the article on her Twitter and Facebook accounts and to academia.edu. It's still night-time on the east coast of the US, but she knows that it's one of the first things they'll see in their feeds when they wake up.

By the time she gets to her shared desk in the department 45 minutes later – no roadworks means a quicker than normal commute on her bike – she's ready to put the rest of her plans to promote her article in motion. She first shoots an email off to the editor of a multi-author blog collective at a top university; after asking how his youngest is taking to school, she inquires if they would be keen for her to write a blog piece based on the article. The blog editor has a pretty quick turnaround, and she knows that there's a chance they might be able to get her blog post up by Wednesday or Thursday if she's quick. Kira also suggests that her publisher might be able to ungate the article for a month or two, so she can link her blog article to the piece. Maybe more than a couple of people might be able to find it and read it that way...

An email pings back from the blog's editor just after lunch. He's keen – when can she send something through? (And his son is doing great at school – though he did get in trouble last week for cutting someone else's hair with safety scissors!) She replies that she should be able to send through her 800-word piece by the end of tomorrow; can he resend the blog's style guide so that she can make sure that she formats the piece to reduce the amount of edits on the other side?

Kira puts together a 'hit list' over lunch of a dozen or so people she thinks will be interested in her new article – some are colleagues and past collaborators who work in universities, there are a couple who work in NGOs in Europe and the US, half a dozen are people working on similar topics that she's met on Twitter, and, if she's honest, there are at least two academic crushes on the list as well. It would be great if everyone could be connected with a tweet, but it's not that simple. Max at Oxfam has given up Facebook for Lent, Shosh at Wisconsin-Madison hasn't managed to get on to Twitter yet, and her adviser's mailbox is bouncing again because it's full. Ted's on a bit of an Instagram binge at the moment – she introduced him to it last month and he's been posting two or three times a day. Thinking for a moment, Kira pulls up free online infographic-maker Infogram and plugs in some of her headline findings. A few minutes later she shares the new infographic on her Instagram, adding the #dataviz hashtag; Ted's seen it and replied within a few minutes. She adds a few more tweets and Facebook updates aimed at those remaining in her Buffer social media scheduling account. Checking her Instagram post,

Kira sees that it's already had a like from *The Guardian's* dataviz team – brilliant! She hopes that they might showcase it on their Data blog. But now, back to those revisions.

Kira spends the rest of the day in the library, finishing off her revisions for her new journal piece. Back at home, she cooks dinner while listening to a podcast hosted by a couple of academics who work in her discipline. The episode is only a week old, so she sends a tweet to the podcast and to the hosts' Twitter accounts suggesting that they check out her new paper. She's hoping that they ask her to be a guest interviewee. After dinner she heads out to meet some friends for a drink locally. Checking her phone on the bus, she sees that one of her previous co-authors – who she did her Master's with and is now at Uppsala – has tweeted her back. She's keen to talk more about her findings, and suggests they take the talk offline. Kira pops her a quick email suggesting a Skype chat Tuesday morning. She knows that talking over the main points of her article will be a great way to get ready to write her blog piece in the afternoon – if she can get her article revisions finished off by then!

*****

We're not all Kira, but if you're a researcher, academic or scientist, then there's a chance you might recognise something from her day as an academic who's plugged into social media. What links all the media and dissemination activities she was involved with in the tale we constructed is that they are all social media, or they show the kinds of opportunities social media can bring to academics and researchers in general.

In this chapter, we'll look at some definitions of social media, including what they are and what they are not; explore their history and how that relates to how social media are used today; and how knowledge organisations have been using social media, and where things now stand. This chapter aims to show any reader – no matter how much or little you use and know about social media – that social media matter, and that they can be incredibly important to your work and career as a knowledge worker. We'll also be giving you an overview of the growth and influence of social media in many different areas of society as a lead in to the discussions in the rest of this book. This book as a whole aims to inspire and energise you as a knowledge worker to do more with social media to share and promote your work, and to use it in accordance with your research lifecycle, much as Kira does in our imaginary example above.

Given the title of this book, the relationship between social media and researchers and knowledge workers is perhaps the most important to us – and to you. Here, rather than box ourselves in, we use the term 'knowledge workers', which loosely means anyone working on research in an organisation. This can include academics in universities, researchers in NGOs, nonprofits and civil society organisations and think tanks, journalists and independent scholars. Social media can be of great benefit for knowledge workers who are also educators, and for how they do research and how this research is promoted.

1.1 WHAT ARE SOCIAL MEDIA?

1.1.1 Defining social media

Ask anyone who works in social media what they actually are and they will probably give you a different definition. There isn't a formal definition, and given the relatively disaggregated nature of online life, there isn't one person or organisation who could legitimately set out a definition that would be universally agreed on in any case. Definitions tend to be based on the centrality of online communities – groups of people interacting online to communicate and share information and ideas. Kietzmann et al. (2011: 241) write that 'Social media employ mobile and web-based technologies to create highly interactive platforms via which individuals and communities share, co-create, discuss, and modify user-generated content'. Similarly, Safko (2009: 6) states that social media 'refers to activities, practices, and behaviors among communities of people who gather online to share information, knowledge, and opinions using conversational media', while Xiang and Gretzel (2010: 180) take more of a consumer-based view that social media 'can be generally understood as Internet-based applications that carry consumer-generated content'. Baym and boyd (2012: 321) argue that, compared to more traditional forms of media, a key feature of social media is its scale:

> It is thus not the ability to use technology toward these objectives that is new with social media, but the scale at which people who never had access to broadcast media are now doing so on an everyday basis and the conscious strategic appropriation of media tools in this process.

Couldry and van Dijck (2015: 2) simply state that 'we side with those who look to resist the redefinition of the social as simply whatever happens "on" social media platforms'. We will be returning to these concepts and exploring the specific characteristics of social media and what they mean for the Research Lifecycle in Chapter 2.

We should also note at this point that throughout this book we will be using the term 'social media' as a shorthand for other digital media, including podcasts, photo- and video-driven platforms such as Instagram, Pinterest, YouTube and Vimeo. The fact that these media are digital means that they can be easily shared and at scale in ways that were only available to professional broadcast media in past decades. Without content, social media networks don't exist – they require text, video and images in order to function as social media – digital media can be this content.

With the definitions of social media appearing to sprawl, in the spirit of Marshall McLuhan's (1994) commentary that the 'medium is the message', it's helpful to go to two of the internet's well recognised – and crowdsourced – spaces for definitions. Wikipedia says that social media are 'computer-mediated tools that allow people or companies to create, share, or exchange information, career interests, ideas, and pictures/videos in

virtual communities and networks' (Wikipedia, 2016). *Urban Dictionary*, a popular dictionary of online slang, says that it is 'Your electronic Second Life' (Urban Dictionary, 2016).
While the latter definition is a bit facetious, it dovetails well with Wikipedia's. In this
book, we see social media as being the ways in which the traditional methods of knowledge exchange have been able to colonise digital infrastructure. Recognising this also
means that we can see social media as acting alongside – not instead of – more traditional
ways of communicating information. Social media just take them further – expanding
their reach and bandwidth. Where knowledge workers could once only exchange ideas
through the spoken word, by letter and printed article, now they also have the options of
the podcast, the tweet and the blog post.

1.1.2   What social media are

At this point, it's also important to outline some of the main social media tools, focusing on those that we'll be discussing more in depth later in this book. Xiang and Gretzel
(2010: 180) write that social media includes 'a variety of applications in the technical
sense which allow consumers to "post", "tag", "digg", or "blog", and so forth, on the
Internet'. As this quote illustrates, there are a large number of social media tools available,
and an even greater number of actions that go with them; two of the four terms that they
refer to are actions that can be performed as well as types of content. We'll focus on the
content rather than the actions for now – though we'll also be going through the actions
as we talk through each type of social media. Figure 1.1 shows some of the examples of
social media we'll be discussing throughout this book. Others, such as Thompson (2013),
have suggested typologies of social media, dividing social media into the public/private
and permanent/ephemeral typologies, but for this stage, we don't need our framework
to be quite this complex (and we'd actually disagree that such clear distinctions can be
made – more on this in Chapter 7).

We divide each of the social media tools in Figure 1.1 into three categories: content,
platforms and tools for collaboration. The *y* axis refers to the potential audience size
that each form of social media can reach, while the *x* axis shows how easily shared each
element is. Tools on the leftmost section are by their nature one-to-one, one-to-few or
few-to-few, with tools approaching 'native' or inbuilt shareability as we move to the right.

Content. This is the 'stuff' that you're trying to gain an audience for. It comes first,
because it's really central to the whole thing. Content is made up of your thoughts, ideas,
reflections and insights. This includes blog posts, discrete chunks of text based on a specific blog platform (of which more in Chapter 3), podcasts (Chapter 5), infographics and
data visualisations (Chapter 4), and images and video (Chapter 6). These are all things
which will more often than not need to be leveraged by social media platforms in order
to get an audience, hence their position midway on the 'shareability' index.

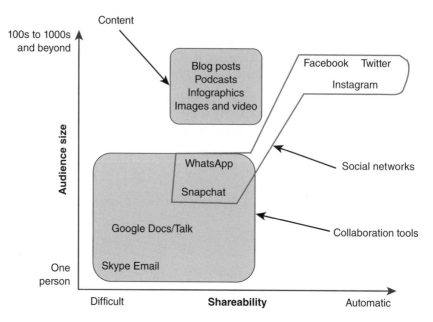

Figure 1.1 Social media shareability and potential audience size

Social networks. These are the platforms that help you get your content to an audience, which can also include your desired audience – though they are not always one and the same. The social networks shown are connective by default – it's their main function and purpose. While their usefulness will obviously vary depending on how linked they have become with others who also use the same network, they facilitate content sharing far more easily than the tools in the bottom left of Figure 1.1. In terms of their potential audience, the sky is really the limit. Follower numbers vary from tens and hundreds to the thousands or even millions.

Collaboration tools. These are the tools of what could be termed the 'old' web. Skype was first released in 2003, and email can be traced back to the earliest days of the internet and was in fairly wide use by the mid-late 1990s. WhatsApp is a mere youngster by comparison, but its antecedents include Google Talk (2005) and even the humble text message, which is also a survivor from the 1990s. The key point about collaboration tools is that they are person-to-person rather than person-to-many. Even tools such as Google Documents only really involve tens or (at most) low hundreds in terms of collaborator numbers.

As you can see from Figure 1.1, there is some degree of category breakdown, and many of these formats are porous and interlinking with some of the others. As well as being useful for collaborative work, WhatsApp and Snapchat are increasingly being used as social networks to distribute content (Morrison, 2016). Most content platforms also have ways of self-distribution – in the case of podcasts, SoundCloud, for example, has its own native distribution network, as does Apple's iTunes. Social networks can also be content. Twitter is famously regarded as a 'micro-blogging' site, and Facebook certainly allows for posts of several hundred words or so.

1.1.3 Social media platforms and their popularity

Now that we've looked at what social media platforms are and how they operate, it's helpful to look at how popular each platform is and where recent growth has been occurring. Table 1.1 gives an idea of just how popular social networks and social media have become.

Table 1.1 Social networks and active users in 2015/2016

Social network	Active users
Facebook	1.65 billion
Tumblr	555 million (Jan 2016) (Statista, 2016)
Instagram	400 million
Twitter	310 million
Google +	300 million (Feb 2016) (Smith, 2016)
Sina Weibo	222 million (Q3 2015) (China, 2015)
Snapchat	100 million (May, 2015) (Tweney, 2015)
LinkedIn	100 million
Pinterest	100 million

Note: Audience figures are from Walters (2016) unless otherwise indicated

1.1.4 What social media are not

The answer to the question of what isn't covered by the term social media is actually very little. While email and Skype aren't traditionally thought of as being forms of 'social' media, they are still ways of sharing information and enabling collaboration with others. In this way they are social – it's just that they speak to a much smaller audience. For us, the important distinction is between social media and social networks (even though the latter are actually a subset of the former). Social networks were created and built around the idea of sharing content with an audience; it's their prime function.

That said, traditional tools like instant messaging applications are becoming more social. Slack, which is a relatively new application, combines the collaborative functionality of an internet messaging service with social network-like functions of mentions and likes. Some websites go even further, publishing their Slack chats as social media content (FiveThirtyEight, 2016).

1.2   THE HISTORY OF SOCIAL MEDIA

To uncover how social media can be useful and important to researchers and knowledge workers, it is useful to spend some time delving into its origins.

1.2.1   From ARPANET to the World Wide Web

Perhaps surprisingly, social media's origins begin with the beginnings of the internet – or what became the internet. In 1958, the US government created the Advanced Projects Research Agency (ARPA) in an effort to counter the Soviet Union's perceived technological superiority following the launch of Sputnik in October 1957 (Hauben and Hauben, 1998). By 1962, the young agency was researching the usefulness of computers in defence and the advancement of technology, an effort which was being led by Dr Joseph Licklider of the Massachusetts Institute of Technology. At this point Licklider actually put forward the idea of an 'Intergalactic Network' – interconnected communities sharing computing resources.

This move towards a community of computing grew at ARPA, leading to the creation of a computer network, ARPANET, in 1969. The network was initially composed of Interface Message Processors (IMPs), located at UCLA, Stanford, the University of California, Santa Barbara, and the University of Utah, but grew relatively quickly, reaching 24 sites by 1972 and 111 in 1977 (Stewart, 2000). Hauben and Hauben (1998) note the close feeling of collaboration among the graduate students who made up the Network Working Group and the military component of ARPA. Robert Braden, who connected the first supercomputer to ARPANET, later stated of his efforts in creating the network: 'The result was to create a community of network researchers who believed strongly that collaboration is more powerful than …. competition among researchers. I don't think any other model would have gotten us where we are today' (Malkin, 1992). The collaborative nature and sharing ethos that characterised the modern world of social media and social networks can be traced back to this early point.

It was with collaboration in mind that email was first introduced into the nascent network in 1972 (Leiner et al., 1997). Email was followed by Listservers in 1975, which allowed users to post 'threads' of comments in response to topics (Preece et al., 2003). Bulletin board systems followed in the late 1970s, allowing early users of home computers to access messages, trade software and play games with people in their town or city (Rafaeli, 1984). In describing the first bulletin board system in 1978, Christensen and Suess (1978: 151) wrote:

> People who left messages saying they had some information of interest and those who said they needed information discovered that other people using the system contacted them. We were pleased to find the system working this way, because that was one of its purposes.

The following decade saw the continued popularity of email, the introduction of Internet Relay Chat (the grandparent of modern messaging systems like WhatsApp) and, in 1991, the birth of the World Wide Web (Preece et al., 2003).

At this point, in the early to mid-1990s the World Wide Web was made up of services like email and newsgroups which users could join and build, but would not actively or automatically connect you with other users. Websites were also mostly one way – information was published with little possibility for interactivity or back and forth.

1.2.2   Web 2.0

The late 1990s saw the beginnings of a movement towards services with user-based and user-created content, with the creation of the earliest social network sites such as Sixdegrees.com and LiveJournal (boyd and Ellison, 2007). These types of service, and the social networks which followed in the early to mid-2000s, became known as Web 2.0 (Van Dijck, 2013). This period saw the rise and fall of a number of social networks, notably Friendster (2002) and MySpace (2003), and most importantly Facebook a year later (boyd and Ellison, 2007). These social networks had the aim of connecting users socially, allowing messages and conversations between users, as well as the posting of updates and photos.

Launched in 2004 on US university campuses, once it became more widely available, Facebook succeeded where earlier social networks had not. This success has been variously attributed to its relatively minimalist approach (compared to MySpace's cluttered anarchy), an initial lack of advertising and the early targeting of college campuses, which had traditionally not been sources of MySpace or Friendster users (Kelleher, 2010; Tsotsis, 2016).

The other major social network that has achieved widespread long-term success, Twitter, was founded in 2006, and actually began as a group text messaging system as an outgrowth of an earlier abandoned podcast platform called Odeo (Arrington, 2006; Carlson, 2011), whereby users could send an SMS message to a shortcode number, which would then be posted online for others to see (Malik, 2006). Unlike Facebook, which introduced photos fairly early on – and can be more closed to those not who are not part of its specific groups and communities – Twitter (at least in its early days) was primarily a text-based update service or 'micro-blogging' platform that was completely open for all to read. As a result, Twitter has often been perceived in the West – especially by journalists – as the method of choice for communication in social movements and protests across the world, though Facebook and other social networks have been used extensively as well. We'll return to this idea in the next section.

1.2.3 Smartphones and apps

It's worth noting that the first smartphone (as we know them now) was launched in the form of the Apple iPhone in June 2007. Smartphones, with their app-based software ecosystem, meant that users no longer needed to be tied to a desktop or laptop computer in order to access social networks. With smartphones, users could update their social media profiles, as well as interact with other users while outside their home or work. Teenagers, who had already been the target of earlier social networks like MySpace, also became significant smartphone users (Lenhart, 2012). By 2014, 74 per cent of US adults were using social network sites, a number which increased to 89 per cent for those aged 18–29. Social media penetration via smartphones is also important; 40 per cent of cellphone users used social media sites on their phone, and 28 per cent did so daily (Pew Research Center, 2013). By early 2015 in the UK, smartphones had become the most popular device for getting online, with 33 per cent of internet users stating that the device was the most important one for getting online (Ofcom, 2015). If we take apps as a measure of the popularity of smartphones, by 2015, there had been 638 billion cumulative app downloads worldwide. This number will only increase; the global revenues from app sales are predicted to rise to US$79 billion in 2020, from US$36 billion in 2015, and app downloads could increase to 378 billion from 211 billion in 2015 (Ovum.com, 2016).

So now that we've taken a look at where social media have come from, and how popular it is, let's look at why it's an important force in the world and what it might mean for your work as a researcher or knowledge worker.

1.3 SOCIAL MEDIA AS AN INFLUENTIAL FORCE IN THE WORLD

The picture we've painted of the rise of social media over the past two decades is one that emphasises the social connectedness that these networks have enabled and promote. But beyond making and maintaining friendships, what is social media good for? Quite a lot, it turns out. The factors which have made social media attractive for social means – low cost and ease of use – also make it useful in nearly all aspects of human life, from politics to business to health.

The overriding feature of the various spheres and examples of the uses of social media that we will discuss in this section is that they are not only a tool for communication and achieving change, they are also tools that can have huge value for building and growing communities with shared interests. These interests could be social, recreational and even political. Social media help support the creation of a 'digital public sphere' (Colleoni et al., 2014: 317) where people with differing opinions and experiences can interact and communicate. Kaplan and Haenlein (2010) cite examples such as YouTube (which they term a 'content community'), which allows users to share video, as well as communities based around sports, such as soccer, on MySpace. For Lovejoy and Saxton (2012), social media tools can be effective at community-building in the nonprofit sphere, commenting that their use can create a dialogue within communities of interest, and that this is 'where true engagement begins, when networks are developed and users can join in the "conversation and provide feedback"' (2012: 350).

It's worth noting, though, that these communities can often become so successful that they begin to form 'echo chambers' (Gilbert et al., 2009), where users reinforce and don't challenge the opinions and views of others. While this tendency can be quite helpful in the formation of social movements (as we'll discuss), over time there is a risk to such communities becoming more insular and exclusive.

1.3.1 Social media and politics: The Arab Spring, #blacklivesmatter and beyond

It's often claimed that social media have acted as a catalyst for change over the last decade (Garst, 2013), though some critics (or cyber-sceptics) are concerned that it promotes what's known as 'clicktivism' or 'slacktivism', where people share or promote an issue within an echo chamber, which leads to little – if any – actual change (Wolfsfeld et al., 2013). That said, the simple act of clicktivism can also be a form of civic engagement itself (Halupka, 2014), does not reduce the likelihood of users being politically active and may in fact make them more likely to engage politically (Lee and Hsieh, 2013; Vaccari et al., 2015). Social media networks can also help to combine real and virtual volunteer networks in times of crisis (Reuter et al., 2013).

Rather than characterising social media as a 'catalyst' for change – which implies that it is the sole or main impetus for change – we can say that social media and social networks can help to promote change, often going hand in hand with (or even amplifying) existing movements or trends (Gerbaudo, 2012; Joseph, 2012). Studying social movements in Chile, the US and the UK, Sajuria et al. (2015) find that social media movements have a tendency to recreate structures of social capital that are seen offline.

Perhaps the most well-known instance of social media playing a major role in world events is during the Arab Spring of 2010 onwards (see Figure 1.2) – though there is significant disagreement as to the role and influence actually played (Wolfsfeld et al., 2013). Beginning in late 2010 in Tunisia, the Arab Spring, briefly, was a series of organised political protests, which led to the overthrowing of a number of governments in North Africa and the Middle East, including in Tunisia, Yemen, Egypt and Libya, protests in Lebanon, Saudi Arabia and Sudan, and with others such as Syria and Libya still consumed in civil war (Lotan et al., 2011).

Figure 1.2 Protest sign using Twitter hashtag during Egyptian revolution, January 2012

Credit: Ahmad Hammoud via Flickr (CC-BY2.0)

Khondker (2011) points out that social media were one of many factors in the various revolutions; their role was critical given that many countries lacked an open media and civil society because of repression by the government. As such, social media played a simple but incredibly important role in the protests that marked the beginnings of the Arab Spring, with activists making use of Twitter and hashtags, blogs and image and video sharing to organise demonstrations and exchange information. Gerbaudo (2012) characterises the digital activists who used social media in this context as 'choreographers', who acted as catalysts for mass mobilisation. Social media were also used by protesters to share stories of what was happening on the ground with the wider world, before Western journalists were able to arrive (Lotan et al., 2011; Murthy, 2013).

In the US, the #blacklivesmatter movement was formed following the acquittal of George Zimmerman in the shooting of an unarmed black teenager, Trayvon Martin. The movement, the name of which is actually based on a Twitter hashtag (Guynn, 2015), gained strength following public anger at further killings of young black men by police in 2014 and 2015 (see Figure 1.3). Harris (2015: 35) contrasts the use of social media by #blacklivesmatter activists with the 1960s civil rights movement:

> The movement's use of technology to mobilize hundreds of thousands of people through social media is light years away from the labor that was once required to mobilize black people and their allies during the 1960s or even a few years ago. Jo Ann Robinson of the all-black Women's Political Council in Montgomery, for instance, spent hours using

a hand-driven mimeograph machine to crank out over 52,000 leaflets that announced a mass protest after Rosa Parks's arrest in 1955. Today, social media – particularly Twitter – can reach individuals throughout the nation and across the world in milliseconds, drastically slashing the time it takes to organize protests.

Figure 1.3 Protesters using the #blacklivesmatter hashtag, in solidarity with Ferguson, MO, encouraging a boycott of Black Friday consumerism in NYC, November 2014

Credit: The All-Nite Images via Flickr (CC-BY-SA-2.0)

While the Arab Spring and the #blacklivesmatter movement demonstrate that by facilitating the creation and building of communities of interest social media can aid social movements in affecting change at a national level, which affects the lives of millions, it can also be effective in more targeted ways by those who want to see more specific changes (rather than the overthrow of an entire government or change at a societal level).

In 2013 a social media campaign launched by the London School of Economics student Caroline Criado-Perez aimed to address the lack of women on UK banknotes, other than the Queen. Her campaign to add the author Jane Austen to the £10 banknote was successful in drawing over 30,000 signatures on the online platform Change.org and convinced the Bank of England to introduce the new note design (Crawford and Gillespie, 2014; Criado-Perez, 2015). It is also important to note that this campaign – and others like it – did result in an online backlash against Criado-Perez and one of her supporters, UK MP Stella Creasy, which included threats of rape and murder. This abuse did not deter the campaign, with Criado-Perez stating on Twitter at the time that she 'won't be silenced by anyone'. In 2014, two people were arrested and convicted for threatening Criado-Perez on social media, with both receiving custodial sentences (Lewis, 2014).

Many political campaigners have taken to social media as a way to engage with the public and potential voters. Most famously, the role of social media in politics gained currency with the then Senator Barack Obama's 2008 campaign for the presidency, which made extensive use of social media tools, mobilising an online network of over 5 million volunteers, and drove over 3 million campaign donations to the candidate (Harfoush, 2009; Kenski et al., 2010; Cogburn and Espinoza-Vasquez, 2011).

With greater youth take-up of social media, political campaigners have been especially keen to make use of it to tap a group which has tended to be less likely to vote (Hendricks and Frye, 2011). Gainous and Wagner (2014) characterise scholarly reactions to social media and politics in two ways: as a way of levelling the playing field between citizens and politicians, or as yet another method of political communication, leading to little real change in the relationship between the governors and the governed.

What is becoming more certain is that social media are an effective tool for influencing election outcomes. Recent research on US online campaigning has found that campaigners' online messages are reaching similar audiences to offline communications, and that they have a positive impact on voter turnout (Aldrich et al., 2015). Similarly, examining more than 100 million Facebook updates, Settle et al. (2015) find that users who lived in US states with more contested elections were more likely to post status updates about politics and were also 40 per cent more likely to vote.

1.3.2   Social media and campaigning: Nonprofits to human rights to health

Social media can also be incredibly helpful for nonprofit and charity organisations, which can use them for advocacy and public education, fundraising and communications with stakeholders. Guo and Saxton (2013) put forward the idea of a mobilisation-driven, relationship-building framework, where nonprofits first reach out and build awareness of their cause among the public, followed by sustaining communities of interest and supporter networks, which is followed by mobilisation and calls to action messages sent to supporters in order to further the organisation's purpose. In their study of nonprofit public relations (PR) practitioners, Curtis et al. (2010) found that those organisations with specified PR departments were more likely to adopt social media.

Joseph (2012: 153) argues that social media also expands awareness of human rights abuses beyond the sphere of the traditional media, bypassing the 'veil of secrecy' that repressive regimes often hold up, contrasting the example of reports from a relatively small number of activists in Syria of the government's attacks on unarmed protesters in

2011 with a massacre of tens of thousands in the Syrian town of Hama in 1982, which was largely unknown to the rest of the world until much later.

Social media have also proved useful in recent years in the area of public health. For example, health researchers can 'mine' large numbers of users' tweets to provide public-health information in real time. This is especially useful in cases of seasonal outbreaks of disease such as influenza or food poisoning (Dredze, 2012), or even weather-related depression (Yang et al., 2015). Social media data can also provide different information from traditional sources, as patients may be more willing to share behaviours and conditions with their social networks than with their doctors. Social networks such as Twitter are also useful for public health institutions to provide health information to the public or even face-to-face via Skype conversations (Murthy, 2013). On the treatment side, social media can also form the basis of patient support networks, with hospitals facilitating patient social media groups using a variety of tools, such as blogs, video chats and Twitter and Facebook profiles (Hawn, 2009). These tools mean that patients are better able to manage their own care, reducing the need for interventions from physicians, as well as support and encourage those with similar conditions.

1.3.3   Social media and business: Profits, losses and lives

Recent years have seen businesses from the largest multinationals to the smallest family-owned shops take up social media to varying degrees and with varying success (Long, 2011). Social media networks, especially Twitter, allow companies to have a direct line of communication to their customers in ways that were either previously not possible or very expensive. Marshall et al. (2012: 357) go as far as to say that social media and related technology provide 'a revolutionary change in the way contemporary selling is conducted' by changing the traditional relationships between customers/clients and salespersons. Similarly, Rapp et al. (2013) suggest that social media have a 'contagion' effect where their use has a positive effect on brand and retailer performance, and increases customer loyalty. Social media use can also lead to more customer visits and greater firm profits (Rishika et al., 2013).

Companies are also able to build up communities around brands and products in a fashion not dissimilar from the patient support networks described above. Corporate engagement via social media can also become incredibly important in times of corporate crisis, and must be managed very carefully; engagement can backfire quickly, such as during BP's Deepwater Horizon oil spill in the Gulf of Mexico in 2010. The corporation was slow to respond to fake Twitter accounts highjacking and impersonating its corporate persona, and later struggled to build up its own social media response, further adding to public opprobrium towards the company (Wauters, 2010; Long, 2011).

With many corporates keen to use social media to promote their corporate social responsibility initiatives, Lyon and Montgomery (2013) argue that the transparency that social media bring may actually lead to a reduction in corporate 'greenwashing' if citizens and activists become concerned that the company is promoting itself to too great an extent.

Now that we've covered how social media can be an important influence across various aspects of society, we're going to look more closely at how social media are used by knowledge workers in education, research and in what's called 'digital scholarship'. We'll see that some of the ideas and considerations we've looked at in this section are also important when working with research and in knowledge dissemination and public engagement.

1.4   SOCIAL MEDIA IN EDUCATION, RESEARCH AND DIGITAL SCHOLARSHIP

The chances are that you have at some point in your working life, either at a conference, departmental meeting or through an internal strategy document, come across arguments for the use of social media in higher education and the research environment. There are now plenty of guides and introductions to using social media platforms too, including our guide to 'Using Twitter in university research, teaching and impact activities' (Mollett et al., 2011), which continues to get thousands of downloads each year, and Bik and Goldstein's 'An introduction to social media for scientists' (2013). Discussions and guides like these centre around the new opportunities and challenges that social media bring to the research environment, and reflect the fact that an increasing number of researchers and academics are using social and digital media in a variety of ways.

While the opportunities for social media may seem apparent, the degree to which scientists and researchers are actually using these platforms varies across countries and disciplines. According to Bastow et al.'s 2012 dataset of UK social science academics, one in six academics (16 per cent) use Twitter (Bastow et al., 2014: 230). A 2014 survey by *Nature* suggests a modest incremental increase, with 25 per cent of the 480 social science, arts and humanities researchers sampled using Twitter regularly (Van Noorden, 2014). Among science and engineering fields, there was a stronger emphasis on academic-focused social networks (Mendeley, Academia.edu, ResearchGate), with nearly 50 per cent of the 3,000 scientists reporting that they regularly use ResearchGate, termed 'Facebook for scientists' by its founder Ijad Madisch (Knapp, 2012). Survey findings of course offer only a limited understanding of usage (not least because a primary way of advertising surveys now is through social media and online channels), but at this point it seems fair to conclude that social media usage rates are strong and continuing to grow in academia, and the same pattern can be observed when we look around the NGO, think tank and charity sectors.

But growing usage figures ultimately provide only a tiny slice of what is going on here. Alongside other dramatic changes to the research environment, like the internet and open access to research publications, social media as a whole remain under-theorised aspects of the research environment. A common conception of social media is that they are primarily used for procrastination (see Riemer et al., 2010; Tervakari et al., 2012; Vanwynsberghe et al., 2015). Alongside research practice, we recognise that it is becoming more common for social media to be used in higher education as part of marketing campaigns (Reuben, 2008) and also as a pedagogical tool (Alexander, 2006; Franklin and Van Harmelen, 2007; Moran et al., 2011).

Digital sociologist Mark Carrigan's *Social Media for Academics* (2016) is an instructional primer on why academics are currently using social media and how they can deal with and address the challenges of social media. The thread running through all these different findings and applications is that there is now a critical mass of individuals using social media for education and research, and what was once recognised as potential opportunities for social media are now becoming a taken-for-granted reality. As Carrigan (2016) notes, social media are facilitating the reach of new audiences, enabling more enhanced collaboration, and are also presenting new challenges for the research community to deal with.

The growing adoption of digital tools and technology in this environment can be broadly understood as 'digital scholarship'. This is a concept that will be explored and interrogated in greater detail in Chapter 2, but for now we will look at how researchers and educators are integrating social and digital media into their working practices and take a brief look at the landscape for digital scholarship.

1.4.1   Social media in practice in the research environment

In an extensive survey conducted over nine months, with more than 20,000 responses, Bianca Kramer and Jeroen Bosman (2015) have substantial data informing an in-depth look at today's research environment. They explore the patterns and process of digital scholarship through scholars' workflow and use of tools. The responses and working practices reported suggest researchers are already integrating a variety of social tools and social media for research and scholarly communication purposes. In Table 1.2, Kramer and Bosman have categorised the developments across six identified phases of the research workflow: discovery, analysis, writing, publication, outreach and assessment.

Current tools and workflows certainly play an important role in connecting research and researchers with the wider society, but even more important for digital scholarship is designing systems that look to align the particular tools with the practices and aims of scholarship (Burdick and Willis, 2011).

Table 1.2 101 innovations in scholarly communication – the changing research workflow

Most important developments in six research workflow phases

	Discovery	Analysis	Writing	Publication	Outreach	Assessment
Trends	social discovery tools	datadriven & crowdsourced science	collaborative online writing	Open Access & data publication	scholarly social media	article level (alt) metrics
Expectations	growing importance of data discovery	more online analysis tools	more integration with publication & assessment tools	more use of 'publish firsts, judge later'	use of altmetrics for monitoring outreach	more open and post-publication peer review
Uncertainties	support for full-text search and text mining	willingness to share in analysis phase	acceptance of collaborative online writing	effect of journal/publisher status	requirements of funders & institutions	who pays for costly qualitative assessment?
Opportunities	discovery based on aggregated OA full text	open labnotes	semantic tagging while writing/citing	reader-side paper formatting	using repositories for institutional visibility	using author, publication and affiliation IDs
Challenges	real semantic search (concepts & relations)	reproducibility	safety/privacy of online writing	globalisation of publishing/access standards	making outreach a two-way discussion	quality of measuring tools
Most important long-term development	multidisciplinary + citation-enhanced databases	collaboration + data-driven	online writing platforms	Open Access	more & better connected researcher profiles	importance of societal relevance + non-publication contributions
Potentially most disruptive development	semantic/concept search + contextual/social recommendations	open science	collaborative writing + integration with publishing	circumventing traditional publishers	public access to research findings, also for agenda setting	moving away from simple quantitative indicators

Credit: Kramer and Bosman (2015), Figshare (CC BY-SA)

As such, at this point, it is helpful for us to go into a bit more detail about what digital scholarship actually looks like in an academic context. As we described in the Introduction, in reference to how LSE has promoted and managed its public engagement in the digital sphere, this can often mean a 'blogs first' strategy, with institutional or individual blog platforms providing the source for content to be distributed via social media. Expert commentary by academics in the UK can be divided into four rough areas: university-based blog sites, externally hosted blog collectives, 'professional' blogs and individual blogs. We will go into these in more detail in Chapter 3, but here are a few (non-exhaustive) examples:

- **University-based/hosted blog sites**: Oxford Politics (OxPol), Ballots and Bullets (Nottingham), Eastminster (University of East Anglia), Edinburgh Politics and IR Blog, Policy Wonkers (King's), Policy @Manchester, the LSE's blogs.
- **Externally hosted blog collectives**: Open Democracy.
- **'Professional' blog/journalism collectives**: *The Economist, The Conversation, The Guardian*.
- **Individual blogs**: Mainly Macro, Enlightened Economist, Tax Research UK, countless other examples.

Before moving on, it's worth briefly focusing on what we term 'professional' blog collectives. Sites such as *The Guardian* and *The Economist* are online versions or representations of what would have been traditional media commentary, which often includes academic commentary. *The Conversation*, on the other hand, is a site for professional academics. It employs journalists to commission academic commentary on current issues, which is then edited with the aim of disseminating it across online media and news organisations. The range of examples of internally-managed and externally-supported platforms with which academics can now be involved is indicative of the growing opportunities for public-facing, digital research content.

1.4.2   Social media in practice in teaching and learning

When knowledge workers also have a role as educators, social media can have a large influence on how students work within their educational environment, by providing a new learning conduit by promoting engagement with content and allowing greater collaboration between students, providing educators are consistent and committed to this engagement (Yaros, 2011; Tess, 2013). Social media may also be part of a 'new way of learning', with less emphasis on individualised instruction and a greater focus on collective and collaborative understanding and exploration (Selwyn, 2012), though some are concerned that the 'conviviality' of social media may be counterproductive in certain educational contexts (Friesen and Lowe, 2012). These new ways of learning can also benefit those who have previously been more likely to have been disadvantaged by older models of teaching, such as students from low-income backgrounds, by building connections with other students and faculty and providing new sources of information (Davis et al., 2012).

At this point it's worth remembering that the internet – and by extension social media – was created by researchers who felt that collaboration was more powerful than competition. Remember those graduate students working with the military in the early days of ARPANET? Social media, by their very nature, encourage this sort of collaboration and the circulation of knowledge. And compared to the pre-digital age, this now occurs far more rapidly (Quinnell, 2011; Beer, 2013). Social media can also provide new opportunities to engage with the participants of research, as well as with those who are affected by it (Durose and Tonkiss, 2013). Where traditionally scholars were engaged in a very straight, unidirectional research process of asking questions, doing research and then reporting results, social media allow us to be more engaged with the subjects of our research and provide new avenues for collaboration with other knowledge workers and inspiration for new research. In many respects, social media 'beg us to reconceptualize what it means to be a scholar and do scholarship in the 21st century' (Sugimoto, 2016). We'll go into more detail about the interplay between social media and academic research in the next chapter.

1.4.3   Social media in practice in think tanks and research bodies

Social media can be tremendously useful in promoting the activities and research outcomes of knowledge workers. While we will go into this extensively later in the book, we'll give a short overview here of how different knowledge organisations make use of it.

We would argue that if research is worth doing (and funding), then it's also worth promoting and disseminating to the wider public. For some knowledge organisations, such as think tanks like the UK-based Institute for Public Policy Research (IPPR), the US-based Brookings Institute and the Urban Institute, for example, undertaking and then promoting research is part of their 'core business'. The IPPR has its own journal, *Juncture*, where it publishes its research and commentary, while Brookings regularly releases books and research papers based on its research. The Urban Institute, based in Washington DC, has its own blog, *The Urban Wire*, which has the tagline 'The voices of Urban Institute's researchers and staff'. Organisations such as these now make extensive use of social media to push their research in support of their agendas (Stieglitz and Dang-Xuan, 2013). These types of organisation are more instrumental in their objectives; while they will have in-house researchers (or commission research) who will be pursuing different projects, these will all tend to be focused in one direction, be it curing cancer for a health research charity or tackling urban inequality in the case of a think tank.

While think tanks and NGOs will tend to have in-house communications teams and motivations akin to those of nonprofits, discussed above, knowledge workers who work in educational institutions such as universities are often 'out on their own', both in terms of the research they do and in how it's promoted. This can simultaneously be very freeing and also constraining on academics' abilities to make an impact with their research.

By early 2016, to say that the social media environment for knowledge workers across nearly all disciplines was saturated would be an understatement, with universities, think tanks and NGOs all making extensive use of blogs and most forms of social networks to promote their research and agendas.

1.4.4   Reflections on the individual benefits of social media for researchers

In the past, the main vehicle for academics to achieve impact was via their journal articles. While there is some confusion as to how often academic papers are actually cited – with a widely held (though likely inaccurate) belief in academic circles that the number could be as high as 90 per cent (Remler, 2014) – it's clear that academic works in journal articles faces a barrier to its dissemination in the form of journal paywalls (where only those who have paid a subscription fee are able to access content). With this in mind, the impetus for using social media to share information from research becomes clear. Social media and networks allow knowledge workers to share the summaries and findings of their work, as well as important aspects of their research process, with large numbers of other social media users who may be interested in their work.

This impetus has become much more prominent over the last decade. One notable example was the association of the 2012 presidential election with 'Big Data' in a way that previous contests had not been, and what that meant for social media commentary from knowledge workers. Nate Silver, through his now-famous *FiveThirtyEight* blog, was able to predict the presidential election result in all 50 states, in a prediction model which outperformed all others. For what seemed to be the first time, the mainstream media were taking bloggers and online commentators seriously in the area of US politics.

Similarly, one of the most popular US academic blogs is *The Monkey Cage*, which launched in 2007. The blog was bought by the *Washington Post* newspaper in the summer of 2013, and moved from its independent blog platform to the newspaper's website. This move, by a large and popular national newspaper, was another signal that academic commentary of the kind that had previously been the domain of journals and books was entering the mainstream. *The Scholars Strategy Network*, begun in 2009 by Harvard political scientist Theda Skocpol, has a more wide-ranging remit, but it produces its own copy of academic commentary to be repurposed for the wider media, rather than operating as its own dedicated publishing platform.

Promoting one's research or findings on social media and social networks is of course a form of public engagement (a topic we situate in the Research Lifecycle and go into greater detail about in Chapter 2) and is the starting point for knowledge workers who wish to begin to build a public profile. Sasley and Sucharov (2014) argue that combining engagement on social media with moral activism can be appropriate and even necessary in some cases. For scholars and knowledge workers who are often marginalised in an academic context, it can also be a means of both self- and community promotion (Grollman, 2015). The move to digital has meant that what were once niche areas of study can often have their own journals or online presence:

> The explosion of Internet access in the 1990s provided academics with a way of sharing their work outside of the traditional publishing route, and new kinds of journals began to emerge. Think of electronic journals and the open access publishing models those journals helped foster as the do-it-yourself record labels of academia. (McCabe, 2013: 55)

In the 21st century, many academics have seen their roles shifting or augmented. Where previously, academia largely consisted of research, teaching and publishing work in academic journals, knowledge work has extended to include self-promotion. In this context, self-promotion does not just include the promotion of articles and research; it also means participating (though not always) in social media discussions in areas of their research and on current events and general societal trends. Social media allow researchers to be social, both with those who share their interests and with the wider world in general. This naturally creates some tensions in terms of academic freedom and in how knowledge workers create a public profile. Sugimoto (2016) comments that while online activity has become the front of house for academics to show their work, policies to protect scholars are currently lacking in this space. Recent years have seen growing incidences of knowledge workers being scrutinised and even sanctioned for comments made on social media, something that knowledge workers should bear in mind when contemplating their own online public engagement.

Another aspect of public engagement is the so-called 'Sagan effect', referring to the success of American astronomer Carl Sagan as a science populariser, but also to the difficulties he faced as an academic as a result (Martinez-Conde, 2016). Despite hosting the most widely watched public television series in US history, *Cosmos: A Personal Voyage*, in 1980, Sagan was denied tenure at Harvard University as well as membership of the National Academy of Sciences in 1991, due to 'the perception that popular, visible scientists are worse academics than those scientists who do not engage in public discourse' (Martinez-Conde, 2016: 2077). Many will recognise a general understanding in academia that public engagement is seen as light, fluffy and not what an academic should be doing with their time. Carroll (2011) has argued that academics look down on those colleagues who have too high a public profile, mostly due to the belief that public scientists care more about their media presence than about their research. But do scientists and academics who undertake public engagement activities such as proactive media coverage, knowledge exchange, and blogging and social media actually under-perform compared to those who don't publicly engage?

In short, no. Jensen et al. (2008) found that scientists who engaged with society were in fact more active academically than others who did not engage. Inactivity in public engagement actually correlated with lower performance. Bentley and Kyvik (2010) found that scientists with popular publications like books and blog posts also had higher levels of academic publishing as well as a higher academic rank.

The idea – as the 'Sagan effect' aptly illustrates – that public engagement by academics may not always please all people all of the time is an important one. Engaging in and with social media is certainly not a guarantee of an academic or knowledge worker's success, and it's important to think critically about how social media have been developing in the academic context.

1.4.5   Critical reflections on social media in academia and research settings

Before we move on to Chapter 2 to consider the ways in which social media have impacted the research environment and the new framework we propose for understanding how social media can be useful, it is equally important to consider the negative aspects of the increasingly embedded nature of these new media in our lives – research lives and otherwise. Social media tools and technology can help researchers act in more social ways, thereby realising the social potential of research, which has, as of yet, been underdeveloped and underexplored. But given the complexity in this space, the application of digital technology must be done in a sensitive manner – sensitive both to the researchers and the researched.

Saturation and balance

It may well be apparent that digital technologies and social media are everywhere and to an unprecedented degree, a phenomenon which has been referred to as a culture, or cult, of connectivity (see Van Dijck, 2013, for an overview). But to what ends and to whose benefit do these technologies operate? Nick Couldry (2015) argues that there is a wider normative shift taking place in society that is certainly sceptical of the role of social media. The heightened level of scrutiny of the place of social media in our lives has been brought on by two reasons, according to Couldry: the unprecedented media-related saturation, or 'supersaturation', of everyday life, and social media's complicated relationship with generating economic value and propelling capitalism through the monetisation of the data of our lives. He writes:

No one doubts the pleasures and benefits of some aspects of social media – what major innovation in history has had no benefits? This issue is balance, and how we get enough distance from our own embedding in social media to assess that balance. (Couldry, 2015: 1)

His strategy for achieving balance is for researchers in media and communications studies to move away from investigations of social media at large, and rather to focus on 'the type of "social" now being constructed through social media. … What should be the role of media institutions in the construction of the social?' (Couldry, 2015: 1). As we let these new media in, researchers should be aware of the shifting relations that this may entail.

Monetisation of data

Another considerable critique of social media's 'supersaturation' of everyday life relates to the monetisation of data and how this may affect the research environment. For example, Gary Hall (2015) has explored the role of academic social networking site Academia.edu and the parasitic relationship its business model could have with the digital sustainability of academic research. While the user-friendly platform enables academics to upload and share their research outputs (a noble aim that aligns well with other digital scholarship trends), its financial rationale (and the $17 million it has thus far received from venture capitalists) rests instead on the data generated by the sharing. This is not unique to Academia.edu but is a hallmark of many digital for-profit services that academics have come to rely upon. Hall writes:

> for the likes of Google, Twitter and Academia.edu free content is what for-profit technology empires are built on. In this world who gate-keeps access to (and so can extract maximum value from) content is less important, because that access is already free, than who gate-keeps (and so can extract maximum value from) the data generated around the use of that content. (Hall, 2015)

Because of the startup's core commitments to its investors, Academia.edu and other third-party platforms used in Higher Education should always be viewed with a degree of scepticism: academic interests are not a core consideration for these companies, and academics and their valuable data are in danger of being exploited.

Optional or obligation?

This situation becomes all the more complicated in an environment where academics are not only encouraged to be more visible on these networks, but now feel obligated to take part in these platforms, whether through internal disciplinary community pressure or through more top-down institutional pressure. Carl Zimmer, the *New York Times* columnist and science writer, has picked up on this trend in a recent commentary piece for the journal *Cell*, noting the rise of headlines like 'Why scientists should write for the public' and 'Why every lab should tweet' (2016: 1094). Zimmer argues that the shift from science communication and public engagement as a voluntary act to a mandatory one could have harmful consequences. In Zimmer's view, most scientists are not equipped with the skills and training to widely communicate their work and by mandating wider communication, the public will become inundated with science content it is not able to filter adequately, 'drowning in an ocean of things to read, watch, and listen to' (Zimmer, 2016: 1095).

The opinion that the general public is not necessarily served by hearing from scientists in their own words is not one that 'the public' necessarily shares. A 2016 Ipsos Mori survey on public engagement with science, funded by the Wellcome Trust, found 'the majority of the public (63 per cent) say they are interested in hearing directly from scientists about the research they are conducting' (Wellcome Trust, 2016: 50). Zimmer's implicit argument for respecting the mediator's role in science communication may be partially motivated by his own livelihood as a journalist and professional science communicator, but the rising trend he has identified in making researchers' public-facing activities mandatory still raises a number of important issues for the academic community and the balance of academic freedom, public engagement and obligation in scholarly communication. Incidentally, the same survey (Wellcome Trust, 2016) found a very low level of public trust in journalists as mediators of medical research information.

In contrast to Zimmer's view of scientists' engagement as a 21st-century symptom of social media, Cassidy Sugimoto (2016) argues that this obligation for academics to communicate their work is no recent phenomenon. The obligation to engage with the wider public over one's work is as old as the concept of academic freedom, and indeed is inextricably linked to it. Sugimoto has traced the understanding of academic freedom to 1915, though notes that we, as an academic community, have not necessarily recognised this obligation. But with the rise of social media, these tools offer the opportunity for scholars 'to fulfill our scholarly obligations associated with academic freedom' (2016). Sugimoto notes that these opportunities are not without complexity, and mismatched incentives and the lack of appropriate social media policies are potentially hampering wider positive engagement.

New vulnerabilities are emerging for academics who have previously had little interaction with the public. In more traditional interactions with the media, the primary risk researchers reported dealing with was the misrepresentation of their work (see Kevin Burchall's (2015) literature review on 'Factors affecting the public engagement by researchers'). With digital media, the risks seem to have multiplied. Deborah Lupton's (2014) survey on academic use of social media reviews a number of perceived risks associated with online technologies in general that demand particularly urgent attention, such as sexual harassment, racist abuse, hate messages and death threats (see Cottom, 2012; Beard, 2013; Kitchin et al., 2013; Mitchell, 2013). Mark Carrigan's instructional book, *Social Media for Academics*, also highlights these existing risks and adds to it a concern over 'the ways in which universities risk stifling the creative possibilities in their concern to manage the risks to corporate identity' (Carrigan, 2016: 16). Carrigan expresses similar worry to Zimmer about the possibility that these engagement activities could be something academics feel forced to do, as this compulsion would take out the freedom and enjoyment integral to many academics' practice (2016: 68). But as Sugimoto argues, freedom and responsibility to engage are not so easily separated.

Who owns and controls social media. and who's watching?

Two of the main attractions of social media are their relative ease of use and the fact that they are free. While, for many, social media services are thought of as a form of new public utility (boyd, 2010), that they are free implies that the user is giving up something in return. For example, Twitter's terms of service states:

> By submitting, posting or displaying Content on or through the Services, you grant us a worldwide, non-exclusive, royalty-free license (with the right to sublicense) to use, copy, reproduce, process, adapt, modify, publish, transmit, display and distribute such Content in any and all media or distribution methods (now known or later developed). (Twitter, 2016)

While Facebook's states that:

> For content that is covered by intellectual property rights, like photos and videos (IP content), you specifically give us the following permission, subject to your privacy and application settings: you grant us a non-exclusive, transferable, sub-licensable, royalty-free, worldwide license to use any IP content that you post on or in connection with Facebook (IP License). (Facebook, 2016)

While these licences essentially enable providers to distribute your content on their network, they do also imply that you do not own your content, which is something that any social media user – knowledge worker or otherwise – should remember. While, for the most part, social media interact with social media providers without difficulty, there have been instances where the provider has been influenced to remove a user's content in incidences of alleged copyright infringement (Edwards, 2015). In 2015 Twitter was criticised for shutting down Politwoops, a site which archived politicians' embarrassing tweets that had been deleted, meaning that this public record was no longer available (Murdock, 2015).

In July of 2016, Facebook was accused of censoring articles and users who had posted about the killing of a high-profile separatist militant in Kashmir (Doshi, 2016).

It's also worth noting that social media are a new venue for state and corporate surveillance over members of the public (Brown, 2014; Trottier, 2016). This surveillance can result in relatively benign outcomes, such as targeted advertising, or more sinister ones, such as police social media surveillance of the Occupy movement (Fuchs, 2013). While most knowledge workers will likely not encounter the latter, more problematic type of surveillance, it's useful to keep in mind for those whose work might intersect with activism or advocating for change.

Who gets to participate?

Social media are frequently seen as democratising, allowing those who have had their voices restricted to be heard, as well as enhancing their ability to build communities. But is this really the case? Meraz (2009) finds that traditional media sources – mostly news – do not have as great a hold on agenda setting and influence online as independent bloggers. In their examination of responses to terror attacks in Norway, Enjolras et al. (2013) comment that those who were mobilised on social media tended to be younger and of lower socio-economic status than those who mobilised through more traditional media channels. Xenos et al. (2014) echoed these findings with their study of social media use in Australia, the UK and the US. In section 1.3.1, we mentioned that social media were also seen as a useful tool for levelling the field in terms of political participation. Using the example of Sweden's 2010 national election campaign, Holt et al. (2013) find that social media use increased the attention and engagement of young people in the political process, while Yang and DeHart (2016) find similar effects in the 2012 US elections for online participation.

But we must also be aware that social media may also help to reproduce existing inequalities of voice. In a study of websites, blogs and social media platforms, Schradie (2011) finds that existing class-based inequalities of content production persist. In their study of online activists in the US, Oser et al. (2013) determine that while women and men are likely to participate equally, and young people are highly engaged, the education and income of such activists are similar to their offline counterparts.

1.5   CONCLUSION

The power and relevance of social media are apparent across virtually every realm of public life, and universities and research organisations are certainly not exempt from this. As we have seen, the adoption of social media is pervasive in our research lives and, as is consistent with the adoption of all powerful technologies, this has meant individual benefits and emerging insecurities (Postman, 2011). If anything, we recognise that keeping up with this environment is overwhelming, if not completely exhausting. To confront this complex landscape head-on, the next chapter will take a step back to explore the conceptual underpinnings of research communication today and investigate where social media might fit more systematically in this wider network of mediated communication in the information age. Through descriptions of the research environment, we present a Research Lifecycle Framework for understanding the role of social media and provide snippets in Chapters 3, 4, 5 and 6 of how blogging, data visualisations, podcasts and photo and video formats offer opportunities and challenges for 21st-century research.

1.6 FURTHER READING

Burchall, K. (2015) 'Factors affecting public engagement by researchers.' Available at: https://wellcome.ac.uk/sites/default/files/wtp060036.pdf [Accessed 7 November 2016].
A good overview of the extent to which researchers engage with the public and why.

Carrigan, M. (2016) *Social Media for Academics*. London: Sage.
A practical guide for academics looking for a broad overview of the benefits and challenges of using social media.

Gainous, J. and K.M. Wagner (2013) *Tweeting to Power: The Social Media Revolution in American Politics*. Oxford: Oxford University Press.
A close look at how social media have changed and are changing US politics.

McCabe, B. (2013) 'Publish or perish: Academic publishing confronts its digital future.' *Johns Hopkins Magazine*, 65(3). Available at: http://hub.jhu.edu/magazine/2013/fall/future-of-academic-publishing [Accessed 7 November 2016].
A short discussion of the challenges facing existing models of scholarly communication in the face of social media and the rise of open access journals.

van Dijck, J. (2013) *The Culture of Connectivity: A Critical History of Social Media*. Oxford: Oxford University Press.
A comprehensive history of the evolution and development of social media platforms.

2

SOCIAL MEDIA AND THE RESEARCH LIFECYCLE

At this point we know that a piece of technology can't make people better or worse. Google isn't making us stupider, Facebook isn't making us lonelier. All technology can do is give us new options for how to behave.

– Goldman and Vogt (2015)

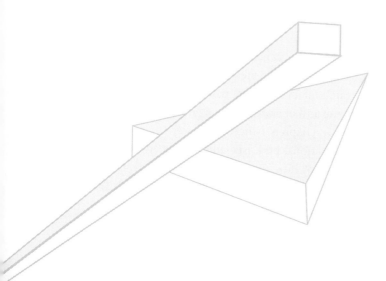

A topic that increasingly affects all of us working in the research community – whether data scientist, anthropologist or research manager – is what the presence and availability of digital media might mean for the research environment as we hurtle through the 21st century. As Chapter 1 laid out, social media and the content which we share through it – such as blogs, podcasts, data visualisations, photos and videos – can be a powerful force in the world, across marketing, business, education, and further. And we have seen that this momentum has led to a complicated relationship to social media for researchers. How can we, as a research community, get to a better understanding of social media's place in the research environment? Can the aims of research and the nature of social media interact in a more cohesive and less conflictual way?

In this chapter, we look at the various conceptual models for understanding the communication of research and argue that due in part to the very influence of the internet and social media on the ways society interacts with knowledge and research, there is a need to reconsider current research practice. For example, a linear model of research communication – where work is researched, published, and then (if you're lucky) disseminated outside academia – does not necessarily serve the interests of researchers and wider society. We argue that the opportunities and challenges of new communication tools actually reflect the changing face of research itself. In this way, social and digital media can be understood less as an external burden placed on the research environment, and more as an available option for a range of modern-day research activities.

In light of the ever-changing models, motives and priorities shaping research communication, we present a Research Lifecycle Framework, which breaks the research process down into six distinct, yet often overlapping, phases: Inspiration, Collaboration, Primary Research, Dissemination, Engagement and Impact. We explore how these phases currently operate in practice and how social media are already being used in support of these activities. We consider what impact this changing research environment might have for how you use social media, blogging, podcasting and other methods of communication to bring to life the work that you are passionate about. Given the limitations of existing models for understanding modern research practice, this framework aims to capture the complexity of the research environment and provides a way to explore how certain media can act in the interests of both researchers and wider society.

With examples and ideas for academics, researchers and communications professionals working in all fields, our Research Lifecycle model will guide you through the changing landscape of how research is done and operates in the world. This chapter is an excellent primer for the following 'how to' chapters which will cover the creation and dissemination of blog posts, data visualisations, podcasts, and photos and videos, and provide further examples and case studies of how the Research Lifecycle Framework is already being applied in practice – but you can also read this chapter afterwards for further contextualisation of your ideas.

2.1 MODELS FOR UNDERSTANDING THE RESEARCH ENVIRONMENT

As we introduced in Chapter 1, if there is a defining narrative for 21st-century research, it is digital scholarship (section 1.4). Social media affords certain opportunities for researchers to explore the research process in more social ways. But what are the theoretical underpinnings of digital scholarship and how does it fit with the wider conceptual understandings of the research process? Furthermore, do these theoretical models adequately reflect research as it is practised today? Here we consider some available models for understanding the research environment in Western developed countries and identify some emerging limitations.

As a preface, we are not suggesting that one model is necessarily right or wrong, and aim instead to look at the variety of conceptual influences that have played a significant role in shaping research and the processes by which it is communicated in society. We are aware that any model has certain advantages and limitations for the research community, and we seek not to reduce the entanglements that exist therein, but instead hope to portray the rich complexity of the modern research environment through these available conceptual windows. To quote statistician George Box, 'Essentially, all models are wrong, but some are useful' (Box and Draper, 1987: 424).

2.1.1 Model one: The linear model of communication

Used by mathematician Claude Shannon and Warren Weaver in 1948 to help explain their work on radio and telephone technologies for Bell Laboratories, the linear model suggests that information starts at one source, flows in one direction and, through a series of steps, is received by another (Shannon, 1948). Typically, this model involves three primary steps: sender, channel, receiver, but can involve many different actors. It can be one-to-one, as in a telephone call, or one-to-many, as in a television broadcast (Shannon, 1948). Beautiful in its simplicity, this model paved the way for the information age. According to historian of science James Gleick, 'It's Shannon whose fingerprints are on every electronic device we own, every computer screen we gaze into, every means of digital communication' (as cited in Roberts, 2016).

For research communication, the linear, transmission model is easily recognisable in the traditional research dissemination process, where the research article is written by a researcher (sender), a publishing outlet is identified (channel) and the research article is published and read by members in the field (receiver). Additionally, many corporate press and communications offices are designed with this linear model in mind, aiming to send research messages via a variety of mass media channels with the aim of reaching a wide net of public receivers.

As an emblematic example of the linear model at work in the research environment, think no further than the influence and reach of Brian Cox, the physicist and Professor of Public Engagement in Science at the University of Manchester (2016). Through a variety of channels (television, radio, books), Professor Cox transmits a range of research ideas through an engaging, visual storytelling format, which is then avidly received by public audiences. His *Wonders of...* BBC television series looked at the Solar System, the Universe, Life and the Human Universe and regularly attracted 3–4 million viewers in the UK (University of Manchester).

Cox's approach to making complex topics accessible to the average public viewer follows in a similar mould to other popular science cheerleaders like Carl Sagan, David Attenborough and the US's Neil deGrasse Tyson. These approaches to research communication require an emphasis on the speaker as primary actor responsible for arranging and presenting the content, and an audience as passive receptors of this information. The other aspect central to a linear approach is the unidirectional flow of information: Cox is communicating his research messages in a clear and powerful visual way, which may well go on to influence people in a number of ways, but the public audience is not able to easily transmit their own thoughts, opinions or disagreements back to Brian Cox in this process.

There are also several limitations to the linear model for research communication. Most glaringly, there is no mechanism for feedback, which is troubling from a moral democratic perspective and from a public engagement perspective (to be discussed further in section 2.2.5). Critics have argued that a linear approach to research communication to solve the 'deficit' of knowledge in society is paternalistic in nature and has also failed on its promise of producing a more informed public (Wynne, 2006; Trench, 2008; Simis et al., 2016). A more pragmatic limitation is that with the unidirectional flow of information, you never know if your communication efforts have ever really been effective. An audience is seen more as an end result rather than a conversing partner. Furthermore, the linear model appears far more relevant and applicable to broadcast-based media. Digital media are not necessarily opposed to the linear model and could be employed to fit within the linear process, yet the opportunities for feedback that social media afford largely exist outside this model.

Despite these shortcomings, the linear model of communication remains a pervasive model for research communication. With ever-increasing demands to reach wider audiences by research funders, it is clear why this linear model is used by many researchers and universities. Fundamentally, the linear model rests on a clear separation between the research and its communication. But is this separation reflected in our current research environment? The UK's Chief Scientist Mark Walport certainly contradicts this understanding, arguing rather, 'Science is not finished until it's communicated' (Yeo, 2013). With the rise of the so-called impact agenda in the UK, where the impact of research is now a core factor in government-funded research assessment, the boundaries between research and its communication overlap considerably (Lewin et al., 2012). For these reasons, researchers may require further mechanisms that more effectively enable two-way exchange. Given these priorities, the linear model appears limited in its ability to capture the communication processes of the modern research environment.

2.1.2   Model two: The convergence model of communication

Looking to address some of the feedback shortcomings of the linear model, the more interactive, convergence model has gained influence over the years and is especially recognisable in the information age. This model was first proposed by social scientist D. Lawrence Kincaid in 1979 and was elaborated on in his book with Everett Rogers, *Communication Networks: Toward a New Paradigm for Research* (Rogers and Kincaid, 1981). Similar to the linear model, the convergence model involves senders, channels and receivers, but rather than viewing communication as a single event, communication acts more like a continuous process. Kincaid summarises the distinction and the more connected interactions:

> The new perspective calls for a paradigm in which communication is treated as a process that unfolds over time and which focuses on the mutual relationships between participants (especially groups of participants), rather than on what one individual does to another individual or mass audience. (Kincaid, 1985: 90)

Thus, in the convergence model, participants create and share information through this process of exchange with one another in order to reach a mutual understanding. With the many channels that currently exist in our digital environment, the convergence model appears more relevant to these types of network and interaction.

The convergence model of communication is highly visible in practice in recent years and particularly among public health professionals who have looked to transmit health information in more interactive ways to improve health outcomes. In 1998, the US Department of Health and Human Services convened a Science Panel on Interactive Communication and Health to investigate the nature of interactive approaches to communication. It identified six key advantages of new media for these aims, including tailored messaging, more flexible learning formats and more convenient access to information across times and locations (Robinson et al., 1998). In Australia, Fotheringham et al. (2000) also confirm these advantages and elaborate on the specific benefits of internet-based technologies in enabling interactive feedback processes for preventive health outcomes.

But there are also limitations to the convergence model's application for research. The convergence model does not deal explicitly with the level of feedback one receives, nor does it provide much clarity in negotiating the different positions of power that exist in society that can shape the type of feedback one receives. For example, Twitter may be an excellent mechanism for two-way exchange, but the type of feedback you can receive varies enormously depending on the individual network. Online harassment and abuse are a type of feedback, although most would recognise this as unhelpful, at the very least, if not violent and harmful communication. Just because there is a mechanism for dialogue does not mean there is a true exchange.

In terms of its implications for research and society, both the dissemination-focused linear model and the dialogue-based convergence model rely on a similar assumption that recognises a binary between research on one side and society on the other. As conceptual categories, this split may appear easily distinct, but in reality it can prove much more blurry. To address this blurriness, a co-existence strategy that recognises the diversity and complexity of research communication in society may be more suitable (Trench, 2008). Rather than a one-size-fits-all approach, these ongoing developments suggest that there are appropriate times and places for different and even occasionally opposing models to research communication.

2.1.3   Model three: The Digital Scholar

A significant limitation of the linear and convergence models of communication is that they do not take as their starting point the situation and working practices of scholars. These information communication models are external to the ideals, practices and processes of scholarship. As an alternative that takes as its starting point a conceptual theory of scholarship, Martin Weller's *The Digital Scholar: How Technology is Transforming Scholarly Practice* (2011) provides a solid basis to understand the gradual yet fundamental shift in academic practice towards the digital. Drawing on Ernest Boyer's inclusive view of what it means to be a scholar, Weller identifies how the four activities of Boyer's model – discovery, integration, application and teaching – are changing with the advent and adoption of new technologies (Boyer, 1997; Weller, 2011; see Figure 2.1).

A brief look at each of the four components, as described by Weller (2011: 41–51), is required:

Discovery: The discovery of new knowledge in a specific discipline or area, what is often termed 'genesis research'.

Integration: Where the discoveries of others are put into context and applied to wider problems.

Application: Related to the concept of service – tackling practical problems and engaging with the wider world outside academia.

Teaching: 'the work of the professor becomes consequential only as it is understood by others'. (Boyer, as quoted in Weller, 2011: 43)

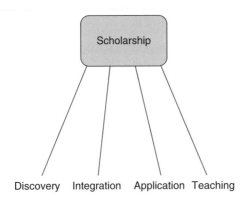

Figure 2.1 The four components of scholarship (according to Boyer, 1997)

Weller's approach to digital scholarship through the lens of Boyer's guiding principles helps us understand the ways technological advances are feeding through an already changing scholarly environment, an environment that is much more connected with and embedded within society:

> The sort of changes we are seeing around open access publishing, development of blog communities, use of Twitter at conferences and easy sharing of content are driven not just by their digital nature but by the convergence of the three characteristics of digital, networked and open. (Weller, 2011: 44)

As a practice-led conceptual model, the Digital Scholar approach is a useful way for understanding the values of research and how communication fits within it. It is arguably the model most closely connected to the aims and reality of the modern-day research environment. Rather than imposing a model of communication that requires a change of scholarly practice, the Digital Scholar model recognises that due to the confluence and characteristics of digital technologies, scholarly practice is, in fact, already changing. There are of course limitations associated with this practice-led approach to communication, namely that it depends on a degree of familiarity with digital tools and also a willingness on the part of scholars to embrace the values of openness.

2.2 THE RESEARCH LIFECYCLE: AN INCLUSIVE FRAMEWORK

The models of the research environment presented each fulfil certain functions. As we have seen, each model has emerged as a response to specific contexts, priorities and realities. No model is in itself sufficient to fully capture the complexity of the research environment. Although the Digital Scholar model does appear to us to be more reflective of the modern research environment, there is no need to reduce the complexity of research communication to one model.

Given the many available options for communication in research today, researchers may benefit from a framework for understanding how research is done, how research operates in the world and, more specifically, where social media might fit in this process. Social media are an opportunity for researchers to behave in more social ways, which in turn can help connect research with its core social remit and obligations.

Becoming more aware of the aims and contexts of our own research settings will help us all in the research community get to a better understanding of which specific media and content will help us in our working practice. We propose that a more appropriate framework is a Research Lifecycle approach, through which research can explore its social underpinnings, reach wider audiences, be used to inspire its users through multiple channels and facilitate much more integrated approaches to research collaboration, engagement and impact.

It is clear that transformations are taking place on a number of levels in academia and we must be careful to not reduce the entirety of this changing landscape to one or two social media platforms. We are particularly wary that rhetoric on social media can often easily veer into unhelpful 'techno-solutionism' (Morozov, 2013) or 'hype'-driven notions of saviourism (Watters, 2013). It would be overly simplistic to argue that how academics communicate with the world is entirely broken and social media are here to fix it. Rather than viewing social media as new transformative technologies that exist to solve some crisis facing research and higher education as a whole, we argue that social media afford certain opportunities for researchers to explore the research process in more social ways, which may in turn help to connect research with its core social remit and commitment, but which also may involve new vulnerabilities. Research, we argue throughout this chapter, is fundamentally a social process. The framework we propose looks at communicating the research process in its entirety but is focused on how social media can be used for communication aims, rather than on how social media are being used in innovative ways to conduct research itself. Certainly, social media are more than just a communication tool – they are now also a very powerful research tool. But this area is beyond the scope of our investigations here. We list some resources on using social media to conduct research in the further reading section at the end of this chapter.

We seek to build on Weller's framework (and, in particular, his Chapter 5 – 'Researchers and new technology') by exploring in more depth the research process as it relates to social media. Since Weller's book was published in 2011, the research environment and digital media have evolved in a number of ways and in particular with an emphasis (for better or worse) on research impact. In light of these developments, we propose a slightly elaborated way of exploring Boyer's notion of scholarship. While Boyer's framework captures scholarship in its broadest sense, we look more specifically to clarify what discovery, integration and application look like as we approach 2020 – leaving aside teaching from our framework as the development of digital media for teaching and learning is beyond our scope and expertise (see Peppler and Solomou, 2011; Benson, 2014; Poore, 2015). We break down each of the following arms of Boyer's ideal framework of scholarship in a way that also captures the research workflow as it exists according to Kramer and Bosman's (2015) summary of digital tools and technologies, discussed in Chapter 1, section 1.4.

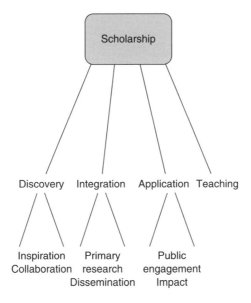

Figure 2.2   Boyer's model elaborated

Credit: Based on Boyer's model, now elaborated, which aligns with Kramer and Bosman's (2015) empirical research on research workflow

It is our hope that the Research Lifecycle model will elucidate the different activities going on within the broad understanding of digital scholarship, grounding Boyer's ideals with current academic practice. Discovery entails the processes of both inspiration and collaboration, Integration entails both primary research and its dissemination, and Application entails public engagement/outreach and impact. These six phases are distinct but often overlap and are not meant to be firm categories. For example, the inclusion of primary research as part of Integration rather than Discovery may sit uneasy with a strict definition of Discovery as 'generative research'. However, modern research has a far greater emphasis on its integration and synthesis with wider interdisciplinary bodies of knowledge.

Given the particular nuances and requirements of the modern-day research environment, we argue that social media are meeting very particular research needs in very particular ways. Being aware of these social needs for your own research project will help you decide where to invest your energies and what types of digital content may be worth experimenting with. As we will explore in greater detail, the research process has evolved in ways that recognise and require a more social emphasis and, we argue, so too should its media.

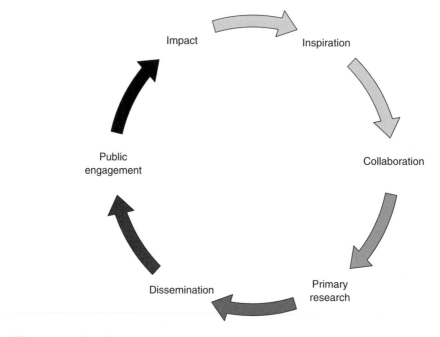

Figure 2.3 The Research Lifecycle

Drawing from the elaborated model in Figure 2.2, and noting the often overlapping nature of these phases, we present the six phases as a cycle that makes up the modern-day Research Lifecycle: Inspiration, Collaboration, Primary Research, Public Engagement, Dissemination and Impact (Figure 2.3).

In the following sections of this chapter, we explore each of these activities individually, providing a description of what the phase is, how it works in practice and how communication through social and digital media might be used to achieve each activity's primary function. We also look to signpost examples, case studies and media formats to be discussed in greater detail in the how-to chapters that follow. Fundamentally, this section aims to equip researchers, administrators and those working in the research environment with an understanding of how social and digital activities fit alongside academic practice in a way that helps researchers feel more confident navigating the types of content and media most suitable to their needs. Through exploring the fundamental aspects of the modern Research Lifecycle, it is clear that research is now, and has in many ways always been, a social process.

2.2.1 Inspiration

What is it?

Researchers do not just wake up with a fully formed idea ready to be researched. Ideas have to be crafted and contemplated long before a grant application is written, whatever the area of research, and the initial ideas that go on to form the basis of rigorously tested and validated forms of scholarship need space and time to flourish. Some may identify with Thomas Edison's reported quip that 'genius is one per cent inspiration and ninety-nine per cent perspiration', but these kinds of narratives do little to demystify where and how scholarly ideas first take form.

What does it look like?

Given the strong influence of the rational-linear tradition of the scientific method as well as Karl Popper's influence in favouring science as dispassionate reason, a phenomenon as subjective as inspiration in research has largely been left for scientists either to ignore completely or to figure out for themselves (Loehle, 1990). In *The Logic of Scientific Discovery*, Popper placed emphasis on theory and hypothesis, and explicitly argued that the topic of inspiration should be ignored, at least logically: 'The question how it happens that a new idea occurs to a man … is irrelevant to the logical analysis of scientific method' (Popper, 2002: 7). And despite decades of scientific progress and world-changing findings, there is still a lack of attention given to the place of inspiration.

Bruno Latour (2005) stresses the complexity of the entire scientific effort, including inspiration, throughout his work, and argues that science itself cannot exist without the 'dynamic network' of social practices. But what does this look like in practice? Analysing a database of millions of biomedical article abstracts, Shi et al. (2015) have demonstrated the extent to which scientists are involved in a complex weaving of scientific innovation, discovery and inquiry, which requires considerable interaction between human and non-human actors throughout the process. Through extensive quantitative and textual analysis, they come to the conclusion that 'the networks described by Latour do more than trace the past politics of science: they act as a substrate for future scientific discovery' (Shi et al., 2015: 73). By applying a hypergraph representation to a dataset derived from the National Library of Medicine's MEDLINE abstract database, they uncover the diverse network of scientific 'things' (methods, fields of study, objects of study, co-authors, etc.) that can affect scientific discovery: 'by wandering across a mental map (or mesh) of science, new associations are woven between things, subsequently influencing what can be conceived, investigated and published in the future' (Shi et al., 2015: 73; see Figure 2.4). In this way, the problem-solving nature of scientific discovery is cultivated by this wider and often unacknowledged social network of research.

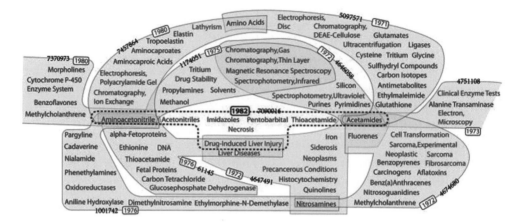

Figure 2.4 Science as a social network: a representation of all the things that can influence scientific inspiration and further discovery in one discreet scientific network

Credit: Shi et al., 2015 (CC BY 4.0)

The idea of science as a social network may provide a good conceptual understanding for research inspiration, but lessons from the 21st century suggest social networks present some particular challenges of their own. Largely because of the fluidity of information across our networks, we are facing an upward battle of dealing with information overload, as Daniel Levitin (2014) has written extensively about in his book *The Organized Mind*. Will social networks help us to manage information and inspiration overload or are they contributing to the problem? He writes:

> Every status update you read on Facebook, every tweet or text message you get from a friend, is competing for resources in your brain with important things like whether to put your savings in stocks or bonds, where you left your passport, or how best to reconcile with a close friend you just had an argument with. (Levitin, 2014)

As Latour (2005) and others have explored, research inspiration occurs across an increasingly complex network of human and non-human actors. But, according to Levitin, given how our brains work on social networks, the network's capacity to actually assist with identifying a research problem and solving it may be diminished by its very nature. The social network of science is not only a highly competitive attention economy within itself, it also exists within wider networks and influences outside the research enterprise. In other words, research inspiration is a highly social process and this is also exactly what makes it so difficult to identify and encourage.

How are social media an essential part of inspiration?

While the challenges to inspiration are apparent given our scarce attention in the age of social and digital media, blogging, podcasts and serendipitous social media platforms also offer particular opportunities for research inspiration. Podcast episodes, for example, often explore an overriding theme from different angles, incorporating a range of disciplinary backgrounds. Whether the podcast is a solo effort or brings together many different voices under a common narrative arc, the format facilitates the weaving of connections across an interrelated set of actors within a wider network. In Chapter 5, we give an example of a podcast series from Douglas McKee and Edward O'Neill at Yale University – Douglas is a lecturer in economics and Edward, a senior instructional designer. Their podcast, *Teach Better*, which they co-host, focuses on pedagogy at the university and features interviews with a range of staff at Yale from the College of Medicine to the Department of Psychology (McKee and O'Neill, 2014). It has opened up a space for colleagues to discuss working practices and gather information across the institution.

Audio media, as separate from other forms of media, adds a level of sensory richness to the research environment that is itself closely connected with the subjective inspiration process. Storytelling podcasts, in particular, can be an immersive, even intimate experience. According to Professor Emma Rodero, a specialist in audio-visual communication, 'Audio is one of the most intimate forms of media because you are constantly building your own images of the story in your mind and you're creating your own production' (as cited in Wen, 2015). The unique way our brains engage with audio media makes podcasts a particularly strong conduit for creation, discovery and inspiration.

Similarly, academic blogging is also proving to be a format that is strengthening the networked ethos. Kaiser and Fecher (2015) show how science blogs can help form new networks beyond traditional disciplines and provide empirical evidence of how academic

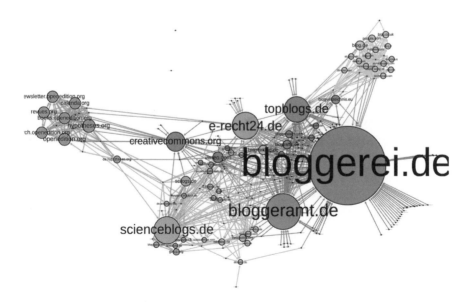

Figure 2.5 A hyperlink analysis of the German blogosphere

Credit: Kaiser and Fecher, 2015. Collapsing Ivory Towers? LSE Impact Blog (CC BY 3.0)

bloggers interact with the wider general blogosphere. Conducting a hyperlink analysis of relationships in the German academic blogosphere (represented in Figure 2.5), they find the current network of interactions in science blogs is leading to a less specialised and more open science landscape.

This analysis suggests that online platforms like blogs are able to facilitate the social network of science, a core aspect of research inspiration and the Research Lifecycle, in new and potentially more effective ways. Chapter 3 goes into far greater detail on how the format and style of blogs provide an effective way to share ideas and pose questions across networks and the benefits this can have for academic research. Blog posts are excellent spaces to introduce ideas that require greater exploration.

So, social media afford new opportunities for finding inspiration in the Research Lifecycle: through blogging, podcasts and the sharing of ideas across a wider connected network. Whether through the intimate experience of audio or through the more informal, question-led content that appears across the academic blogosphere, new formats are opening up wider possibilities to explore the often-overlooked 'ideas phase'.

2.2.2 Collaboration

What is it?

Collaboration is closely linked to the inspiration phase outlined above. Where inspiration is the spark of an idea, collaboration is the next stage of exploration. This typically involves many diverse networks and increasingly different methods, and is often understood as instrumental to research, but with little support given for why, to what extent, or how collaboration works in practice (Katz and Martin, 1997).

What does it look like?

In 2015, a single physics paper by the Large Hadron Collider research team at CERN had over 5,000 authors and broke the record for most contributors listed on a single paper (Castelvecchi, 2015). No, that's not a typo: 5,000 authors. In the age of CERN and other large-scale, team-oriented research projects, the idea of the solitary scientist working alone at their desk seems particularly outdated (if this was ever true) (Stokols et al., 2008). With research also increasingly involving cross-disciplinary partners to respond to society's most pressing issues, collaboration certainly seems to be becoming the new norm, where 'each team is a sprawling collaboration involving researchers from dozens of institutions and countries' (Castelvecchi, 2015). For disciplines in the humanities and social sciences struggling for research funding, interdisciplinary collaboration has been touted as an attractive opportunity for renewed relevance and survival, with funding organisations like Science Europe and the British Academy taking up the interdisciplinary call to arms in recent years.

Often, co-authorship patterns are used as a proxy to demonstrate the extent of collaboration that occurs in research. Larivière et al. (2006: 524) analyse the co-authoring practices of Canadian researchers in the social sciences and humanities compared with those in the natural sciences and engineering:

> In 2002, more than 2 out of 3 Canadian papers in the social sciences had multiple authors, compared to 1 out of 2 papers at the world level. In the humanities, on the other hand, the vast majority of articles were written by just one scholar.

Wagner and Leydesdorff (2005) also examine authorship data from the Science Citation Index and confirm the growth of international collaboration in science. Looking at the implications of their analysis, Wagner and Leydesdorff consider competition as a key reason why co-authorship has continued to strengthen:

> The many individual choices of scientists to collaborate may be motivated by reward structures within science where co-authorships, citations and other forms of professional recognition lead to additional work and reputation in a virtuous circle. Highly visible and productive researchers, able to choose, work with those who are more likely to enhance their productivity and credibility. These 'continuants' mediate the entrance of juniors into this network. This creates a competition for collaborators. (2005)

Humanities authorship, according to Larivière et al. (2006), is still predominately single-authored. Some preliminary analysis by Nyhan and Duke-Williams (2014) finds significant increases in dual- and triple-authored papers from two prominent digital humanities publications, although single-authored publications were still the norm here as well. There are certainly limitations to the proxy of co-authorship for collaboration, not least of which remains the widely recognised disconnect between publication practices and research practices, but as collaboration is an unwieldy process to define, let alone measure, co-authorship data provides a useful starting point.

One aspect that makes collaboration difficult to assess is that it can occur between and across many different levels (Glänzel and Schubert, 2005). It can refer to coordinated behaviour between and among individuals, research groups, departments, institutions and countries. It can refer to formal research activities like data gathering, analysis or grant writing, but can also be used to describe the informal aspects that occur in the day-to-day interactions within and without universities and research centres, which may or may not also go on to influence primary research. Increasingly, collaborative research refers to research activities involving non-academic partners, like community groups and policymakers (see, for example, UC Santa Cruz's Center for Collaborative Research for an Equitable California).

Katz and Martin (1997) have written what is recognised as an extensive overview of what research collaboration is and how it operates in the research environment (Beaver, 2001;

Lee and Bozeman, 2005). The authors outline a number of recognised benefits of research collaboration, which continue to underpin research today. According to Katz and Martin (1997), the benefits of research collaboration are as follows:

1. Modern research is increasingly complex – no one researcher will possess all the knowledge or skills so collaboration is effective job sharing.
2. Skills and knowledge are best transferred by learning on the job by engaging with others.
3. Cross-fertilisation of ideas may generate new insights.
4. Collaboration offers intellectual companionship in an otherwise isolating profession.
5. Collaboration plugs the researcher into a wider network of contacts.
6. Collaboration enhances the visibility of work.

In short, collaboration is a sprawling idea, occurring early in the initial research phase, but also after findings have been discovered and the concept as a whole is perhaps ill-defined. But it may also be beneficial for this to remain vague – research collaboration aims to capture all the social interactions of the research process, informal and formal, and may always elude a narrow output- and outcome-oriented institutional understanding of research.

How are social media an essential part of collaboration?

The role and function of collaboration today suggests many great opportunities and benefits for researchers across disciplines. Researchers are now considering how their working practices can facilitate a more cooperative approach to knowledge production. To what extent are the available tools and media supporting this activity?

With an increased emphasis on collaboration between teams and a strong need to share findings across dispersed and often interdisciplinary colleagues, 'researchers are increasingly relying on computer and communication technologies as an intrinsic part of their everyday research activity' (De Roure et al., 2003). Certainly the growth of internet-enabled communication and web technologies, which can assist with co-authoring, data-sharing and citation sharing, such as Google Docs, GitHub and Zotero, signal a response for research collaboration. But what types of social platform and content might work best for communication aims? Underexplored as yet is the role of social media for these purposes.

The findings from a 2014 survey conducted by Deborah Lupton on perceptions of social media among researchers suggest that social media may continue to play a more substantial role in facilitating research collaboration. Lupton (2014) writes:

> for the majority of respondents, the principal benefit they gained from using social media was related to the connections or networks they had established with other academics, students and also those outside academia. Frequent reference was made to the wide

scope of these connections, which allowed people to interact with others across the globe and from diverse communities. Many respondents referred to the opportunity their social media use gave them to connect to other academics in other countries or in other academic disciplines as well as their own and have stimulating conversations or eliciting feedback on their research online.

These findings on the benefits of social media for academics mirror the benefits more broadly listed as part of research collaboration in general. This is partly down to social media's very nature. According to social media expert danah boyd, the four specific characteristics of social media are that they are persistent, replicable, scalable and searchable (boyd, 2011). Due to these conditions, researchers can now connect with wider networks of people and bodies of knowledge, academic and non-academic alike, which may have previously been made more difficult by gatekeepers or paywalls (Carrigan, 2016). The type of collaboration that researchers are looking to engage in and the specific aims of knowledge production that this collaboration is looking to facilitate will largely dictate the most suitable platform and content.

For example, Chapter 6 on photography and video explores how sociologists at the University of Maryland use the visual bookmarking website and app, Pinterest, to promote the teaching and learning of sociology. Popular with students, professors and the wider public, the visual medium connects users to images and videos that convey sociological concepts in everyday life. Especially for informal and opportunity-driven research collaborations, social media tools can be incredibly effective.

2.2.3 Primary Research process

What is it?

The next phase of the Research Lifecycle that follows from collaboration is the actual fact-finding aspects of research, or primary research. This can entail investigation, testing, exploration and analysis, depending on the type of research being conducted. It is clear from the previous phases of inspiration and collaboration that modern research involves a complex network of partners, so it is not surprising that the research process itself should involve an increasingly complex relationship with vast amounts of available data.

What does it look like?

Indeed, the vast amounts of available data now at researchers' fingertips have presented numerous opportunities for primary research in statistics, biology and genetics, as well as the growth of theoretical research in the social sciences critiquing 'Big Data' methodologies, with research institutes, journals and councils devoted solely to exploring the role of data in society. The era of Big Data may be well underway, but as boyd and Crawford (2012) point out in their consideration of the implications of the Big Data shift, the significance of this transition is more to do with our newfound capacity to 'search, aggregate, and cross-reference large data sets' than the quantity of the actual data. Not only are researchers able to collaborate with each other in unprecedented ways, but our datasets are able to communicate, combine and integrate with equally profound ramifications.

How are researchers keeping up with such a rapid expansion of available data and opportunities for analysis? One way for researchers to cope with the scale has been to enlist volunteers to help with the research process through crowdsourced 'citizen science' initiatives. Crowdsourcing is based on a simple but powerful concept: virtually anyone has the potential to plug in valuable information (Greengard, 2011). And plugging information in has become significantly easier in recent years. Crowdsourcing, particularly in relation to text and data input and analysis, has become a popular feature of Web 2.0, with many industries and organisations tapping 'latent talent' to further specific aims (Greengard, 2011).

The term 'citizen science' has become a buzzword more recently. It refers to large-scale, research-based projects relying on the crowdsourced labour of enthusiasts. Silvertown (2009) traces the concept further back, arguing that all science was citizen science before the professionalisation of science in the late 19th century. But in its more modern form, members of the public have contributed to the collection and analysis of academic research in a number of fields. The most well-known example is the University of Oxford's GalaxyZoo project, which has over 1 million online contributors and encourages members of the public to classify galaxies. But there are also high-profile examples in ecology, such as the British Trust for Ornithology Birdtrack project and iNaturalist, an app that compiles observations of biodiversity across the globe. Crowdsourcing has stretched beyond strictly scientific disciplines, and humanities projects have also recognised the opportunities (Dunn and Hedges, 2012).

Citizen science has been received with positive enthusiasm from the public and scientists alike, and in particular, as Dan McQuillan (2015) has argued, the democratic and authority-challenging aspects of the movement have the potential to 'transform the terms on which science and society meet'. In addition to the cheap research labour that volunteers provide, this partnership approach has also helped boost public interest and engagement in the research, although this is not without issues for the research community, as highlighted in Matsakis (2016), Irani (2015), Matias (2015) and Mims (2010). With research funders increasingly looking for their research to demonstrate public relevance, citizen research initiatives are able to lay the groundwork for an engaged public audience, while also collecting and analysing the research more efficiently through the help of public volunteers.

Archaeology has particularly benefited from citizen science and crowdsourcing in recent years, and these new directions demonstrate just how invigorating new technologies can be for a field. Appearing on BBC Radio 4's *Today* programme, which gets an average of 7 million weekly listeners (Sweney, 2016), Lisa Westcott Wilkins, co-founder and managing director of DigVenture, a crowdsourcing platform that hosts a range of archaeology and cultural heritage projects, underlined the opportunities that working with volunteers has opened up for the discipline: 'Archaeology has traditionally been a closed discipline. But the way we are working now, the enabling factor of tech, means that anyone can really be part of our dig' (*Today*, 2016). She describes the lengths to which the team goes to ensure every aspect of the excavation is made available to the public, whether through live video feeds or by making images of the discovered objects available in real-time or to training volunteers to work on-site. Sarah Parcak, the 2016 TED Prize winner and satellite archaeologist, explained the wisdom and value that crowdsourcing technology offers for the future of archaeology:

> it's not just about finding the things, it's about getting people connected to each other. We have no idea what's going to happen when the crowd gets together. They are going to organise tweetups and go cleanup an archaeology site or go volunteer at a school. We really have no idea what the next generation of kids in the world [are going to do and] who are going to be engaged with archaeology in such a different way. (*Today*, 2016)

There is certainly a degree of insecurity in opening up the research process, and an even greater amount of planning, infrastructure and support required to systematically capture wisdom from the crowd. Involving volunteers, though arguably less problematic than poorly-paid, crowdsourced labour like Amazon's Mechanical Turk, is implicated in a complex and unequal relationship of power between the mass of volunteers gathering the data and the select research team largely deciding the research questions for investigation and primarily receiving credit and recognition for the work. Teams looking to involve external partners in research should all consider how to navigate this complicated landscape and seek to minimise any exploitations. But, as Parcak and Wilkins both attest, engaging with the social elements of research and bringing the crowd in as early as possible offer limitless possibilities for the research itself and the field in general.

How are social media an essential part of the primary research process?

Certainly not all research projects are suited to an app-based communication approach or a live 24–7 Periscope video feed, but a greater awareness of how different publics can at different stages be brought into the primary research phase offers promising directions for the future of research. The power of crowdsourcing has been identified in a number of different industries and integrating a more open, transparent and citizen-focused ethos throughout the research phase is invigorating fields. Every discipline can gain from the wisdom of the crowd and greater dialogue with interested communities.

Photos and video are digital media with great opportunities for primary research and have been in use as effective research methods in disciplines like anthropology for years. Chapter 5 looks in greater depth at several examples of podcast series that have been produced with demonstrable benefits in the primary research phase. We look at how podcasts like *Researcher in the Field* and *LSE Review of Books* are able to explore more personable, face-to-face stories that have traditionally been left out of research methods.

Through crowdsourced, social media engagement and by exploring fieldwork through visually engaging digital formats, research teams can efficiently and creatively proceed with their research. By openly sharing data in real-time, co-producing research with public partners and contributing to transparent conversations about what research is and what findings might mean, social and digital media present numerous benefits for the future of research and society.

2.2.4 Dissemination

Now that the primary research has been produced, it is necessary for researchers to communicate these findings outside the research team. While this directive may appear clear, scholarly communication is no simple process. According to the Association of Research Libraries' definition, scholarly communication is 'the system through which research and other scholarly writings are created, evaluated for quality, disseminated to the scholarly community, and preserved for future use' (ARL, 2016).

What does it look like?

In many ways research dissemination has become an end in itself in the academy. It is perhaps best encapsulated by the phrase 'publish or perish'. But it is not just quantities of dissemination that are important; academic careers also depend on where one publishes their work, as an indicator of wide dissemination. Of course, the process of dissemination does not require formal academic publishing. But due to a complex interplay of research assessment structures, scholarly communication filters and what Cameron Neylon has referred to as the disciplinary 'politics of exclusion', academic publishing via monographs (for humanities and qualitative social sciences) and journal articles (for STEM and quantitative social sciences) has evolved as the primary mechanism of research dissemination (Neylon, 2015).

But this prominent emphasis on academic publishing as research dissemination is not without shortcomings and there has been growing attention in the last 20 years on the negative consequences a print-based model of academic publishing has for the production of knowledge in society. Before the internet, journal publishing was fairly straightforward. Publishers coordinated peer review and the physical printing of relevant scholarly articles in bound collections (journals). Those who were interested in the latest disciplinary findings, made sure to keep up to date with the latest issues. But with the advent of digital publishing and processes, the processes (and certainly the profits) of scholarly publishers have been called into question. In an interview on the future of publishing and how we read, technology writer and general supporter of digital 'disruption', Clay Shirky infamously noted that publishing is no longer a job or an industry, it is a button:

> In ye olden times of 1997, it was difficult and expensive to make things public, and it was easy and cheap to keep things private. Privacy was the default setting. We had a class of people called publishers because it took special professional skill to make words and images visible to the public. Now it doesn't take professional skills. It doesn't take any skills. It takes a Wordpress install. (Shirky, 2012)

Alongside the ease of publishing that the digital age affords, the issue of economics is also at play here. Paywalls and journal subscriptions have been noted to have caused a crisis in scholarly communication: the so-called 'serials crisis', where a growing number of relatively well-off Western research libraries are simply unable to afford the purchasing costs required for their academics to access journal content (Mobley, 1998; McGuigan and Russell, 2008).

Furthermore, huge disparities in knowledge creation and dissemination have been pointed out that negatively affect research and researchers. The map in Figure 2.6 depicts this imbalance, showing territory size according to the proportion of authors of published scientific journal articles. Roughly three times more scientific papers per person living there are published in Western Europe, North America and Japan than in any other region.

It is not only considerations of where you publish but also *from* where you publish that play a defining role in the communication and dissemination of your work today. The use of the discredited Impact Factor as a proxy for impact and quality of scholarship in STEM fields and the more subjective understanding of publisher prestige in social science and humanities disciplines certainly have negative effects for scholars in Western countries (Konkiel et al., 2016), but lead to even more disadvantages for scholars from developing countries (Alperin, 2014). This inequality does not sit well with an academic publishing and dissemination system that purports to be based on quality and rigour.

In response to both the inequality of the research dissemination system and the potential benefits for economic and social development, certain regions have developed their own scholarly-led approaches to dissemination, enabled heavily by current advances in digital technology (Alperin et al., 2012). The most well-known of these are Latin American initiatives, such as SciELO, REDALYC and Latindex, which have established a collection of peer-reviewed, regional journals, free to read and free for authors to publish in. Reflecting on the extent of visibility of research in Latin America, Dominique Babini (2013) writes, 'Today, open access [to research publications] in Latin America is the standard'.

Figure 2.6 Disparities of global knowledge production

Credit: Worldmapper.org / Sasi Group (University of Sheffield) and Mark Newman (University of Michigan)

Note: Territory size shows the proportion of all scientific papers published in 2001 written by authors living there

The Latin American context for open access is far removed from the situation elsewhere, however, and the UK and other Western countries are in many ways still 'catching up' to the opportunities and benefits that this transition could bring for wider dissemination of academic publications (for more on this, see the *LSE Impact Blog* eCollection: Open Access Perspectives in the Humanities and Social Sciences, produced in collaboration with SAGE). Because of this, using dissemination interchangeably with the term 'publishing' may further complicate matters, given the fraught relationship that has emerged. To demonstrate the fundamental principles of research dissemination (and how this has come into conflict with formal academic publishing interests), we briefly consider the context of scholarly file sharing in the digital age.

In 1994, Stevan Harnad, a cognitive scientist from the University of Southampton, sent around a conference abstract via an email listerv proposing a radical new way for sharing scholarly research findings. He argued that for centuries researchers had been forced to make a 'Faustian bargain' with publishers, allowing them to charge for access to their research as this remained the only practical way to make their work public (Okerson and O'Donnell, 1995). But with rapid advances in electronic publication, this no longer remained the case. The dissemination of scientific findings could be done much more efficiently and cheaply than print-bound journals through a public file transfer protocol (FTP) or a standard protocol used to share computer files on a computer network. In practice, this would work by researchers uploading electronic versions of their articles to a globally accessible portal.

Harnad titled his abstract 'A Subversive Proposal', and although the idea was met with excitement, scepticism and dismissal by different parts of the scholarly community, by today's research standards, the idea of finding and sharing an electronic version of a scholarly article is so embedded in daily practice it would hardly merit a label as strong as 'subversive'. The landscape of scholarly services is now largely dependent on the public FTP infrastructure outlined by Harnad, including the existence of institutional repositories, academic social networking sites like Academia.edu, ResearchGate and even the ubiquitous scholarly search engine GoogleScholar. Further underlining the proposal's pervasiveness, researchers are now mandated to provide electronic versions of their articles and conference proceedings by the Higher Education Funding Council for England (HEFCE) if they are to be eligible for research assessment (and, thus, research funding). So how have we got here? To what extent has the move from print to electronic publishing shaped how researchers disseminate their work? And what might this mean for the future of scholarly publishing?

According to Harnad, academic authors' primary motivation for publishing is to get peers and the public (or at least the small portion of the public interested in such 'esoteric' research findings) to see their work. He refers to academic publications as PUBLICations, emphasising the intrinsically social act of scientific publishing (Okerson and O'Donnell, 1995). But he also recognised this transition would not be entirely seamless. He foresaw two complicating aspects which would prevent the overnight transition from print model to 'purely electronic publication'. The first complication was that these print model publishers were still largely responsible for the coordination of peer review and quality control of scholarly articles and, second, the reputational aspects related to the publishing venue where your work appears. Both of these complications could be overcome, he argued, as long as publishers allowed their authors to substitute the peer-reviewed accepted versions for their previously uploaded preprint versions. This would benefit authors and readers as it would provide immediate and global access to a wide range of scholarly content while also ensuring the scholarly record was kept intact. Print-model publishers would want to allow this uploading because, otherwise, their authors, on whom they rely so heavily for content, would simply choose to publish with electronic-only publishers.

In reality, most publishers did not see it this way and sought instead to allocate limited electronic access on their own terms, rather than cede control on the Web. In the 20 plus years since Harnad's subversive proposal was first shared, the academic publishing market as a whole is still coming to terms with the opportunities, challenges and controversies of making scholarly content public in the digital age. Peter Suber, the director of Harvard Library's Office for Scholarly Communication, has written extensively on this period of transition, which he refers to as the 'access revolution' in his book *Open Access* (Suber, 2012). There are a variety of different stakeholders in the scholarly community responsible for addressing academic dissemination in the digital age, including but not limited to academics, research funders, university administrators, librarians, publishers, taxpayers and the wider public, but, at the same time, each has their own set of requirements and preferences for how scholarly material might be made available. Suber (2012) has explored in detail these many dimensions affecting the adoption of wider open access to electronic versions of scholarly material.

Of particular interest for our Research Lifecycle and how dissemination occurs in the digital age is the way that disciplines have responded to this call in different ways. STEM disciplines are largely considered to be further along in the 'access revolution' and have a number of outlets and business models supporting digital innovation in the scholarly publishing market. The social sciences and humanities face practical barriers, like limited funding and other more complicated cultural and institutional barriers (LSE Impact of Social Sciences Blog, 2013). The two fundamental complications Harnad had identified (peer review and journal brand) continue to play a leading role in shaping the debate.

How are social media an essential part of dissemination?

Of all the Research Lifecycle phases, there has been considerable focus on the benefits of social and digital media for the dissemination of research and its role in helping academic research connect with audiences both inside and outside the academy. According to Bastow et al. (2014), the digital age has played a key role in 'disintermediation' of the communication chain, a cutting out of those traditionally relied upon for research to filter through to wider society: 'From the late 2000s onwards, blogs, Facebook, the micro-blogging site Twitter, and other social media (such as Pinterest for images), gave academics a greatly expanded opportunity to undertake their own dissemination activities directly' (2014: 228).

Due to the disruptions of traditional media and the decline in print-based publications, there is a growing audience and appetite for online content. Blogging fills this gap considerably and academic blogging in particular has enjoyed distinct advantages in this landscape and has seen rapid expansion due to the public trust and value in academic voices (Ipsos Mori, 2016). Chapter 3 goes into further detail on the distinct opportunities for academics looking to get more eyeballs on academic work via the medium of a blog. The benefits and varieties of blogging formats are explored at length and, in particular, section 3.3 presents an overview of the advantages blogging can have for dissemination activities related to academic work, such as potential citation benefits to the formal academic publications mentioned in blogs, reaching wider non-academic audiences such as policymakers and civil servants, the freedom to control publication release in real-time, and experimenting with writing in more accessible styles.

More interactive, visual formats are also emerging as popular mechanisms to disseminate information. For example, Chapter 4 looks at the Brazilian think tank FGV's Department of Public Policy Analysis (FGV/DAPP), which has used infographics to garner considerable attention on the dengue epidemic in Brazil. For time-poor online audiences, infographics can be more appealing to look at and more understandable than a long text-based article or research report.

Social media offer some truly new ways and innovative formats for presenting and sharing academic content. Yet so far social media's use for dissemination purposes for academic practice has been overwhelmingly occupied with reaching *academic* audiences more effectively, either through open access publication or text-based communication styles that are similar in content to the traditional journal article, if not in form (Mewburn and Thomson, 2013). And certainly this kind of practice is still worthwhile. But there is also considerable potential for less traditional, more audience-engaging content formats like infographics and video to play a much bigger role in the dissemination phase of the Research Lifecycle.

2.2.5 Public Engagement/outreach

What is it?

The term 'public engagement' is associated with the many ways Higher Education institutes engage the public with their work. The distinctions between dissemination, engagement and impact are tricky and in academic conversation are often used interchangeably. Each of these three phases exists in a complex relationship to the others and there is also significant overlap across these activities. To provide a brief (and ultimately simplistic) explanation for our purposes, *dissemination* is focused on how many people, academic and non-academic, are coming across your work; *engagement* is focused on the conversations, online and in-person, that your dissemination efforts have led to and the two-way exchange that may occur; and *impact* is a demonstrable change that has resulted from either the dissemination or public engagement activities.

The final two components of our Research Lifecycle model, public engagement and impact, pertain to Boyer's idea of 'application' of scholarship (see Figure 2.7). In practice, Boyer's scholarship framework of discovery, integration, application and teaching parallels the logic model trajectory of inputs, outputs and outcomes (Davis et al., 2007; Hofmeyer et al., 2007; He and Jeng, 2016). The previous sections have covered the conditions required to make research (inputs) and how the research itself is presented in a coherent format (outputs). Outreach and public engagement activities can be seen as both output and outcome. This section will explore the connecting role public engagement can play between the previously discussed dissemination function of scholarship and the Research Lifecycle's final component – outcome-oriented impact.

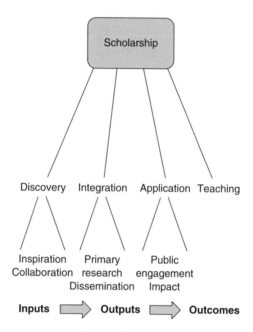

Figure 2.7 Boyer's scholarship, Research Lifecycle and the logic model

What does it look like?

According to TNS-BMRB, an independent social research organisation, eight out of ten (82 per cent) researchers carried out at least one form of public engagement in 2015 (TNS-BMRB and PSI, 2015). In recent years, universities have looked to embed engagement across their core commitments to teaching, research and service (O'Connor et al., 2011). A number of studies look at how researchers' involvement with non-academic stakeholders has shifted over the last 20 years (see trends in integrative research, participatory research and action research, for example).

Bauer (2009) charts the rise of public understanding of science (PUS) as a movement following emphases on improving scientific literacy and the evolution of the term in academic contexts. From as early as the 1980s, ideas regarding the role of science in society were challenged as being more about a public attitudes deficit than a particular literacy deficit (Bauer, 2009). The lack of significant support for science initiatives was argued to be a result of a lack of understanding about the value science and scientists bring to society. With more awareness and communication channels, it followed that scientific concepts would be held in wider esteem. As we discussed in section 2.1, social scientists in particular have been involved in critiquing this deficit model, noting its realist failure in affecting change, and also its condescension of a deficient public (Wynne, 2006; Trench, 2008). Rather, engagement through dialogue with the public (and indeed different publics) was promoted as the antidote to an unreflective understanding of how science functions in and through society.

Overall, the shift from deficit to dialogue, rhetorical or otherwise, has opened up many opportunities for research communication to be more than a one-sided exchange, and instead a continual process of evaluation and scientific critique. As institutional recognition of these mechanisms for engagement has broadened across funders, Higher Education institutes and policymakers, more money has been made available for experimentation with engagement models and practices. Paraphrasing Newton, Ann Grand (2014) describes how new trends towards openness are encouraging wider consideration of the processes by which the public can 'peer over the shoulder of giants'.

Attempts to measure the public engagement with research has, however, been fraught with difficulty. Does this engagement process occur before, during or after the research? Can interactions and conversations be quantified, and when exactly does engagement activity end and research impact occur? Are some disciplines and topics able to be engaged with more than others? These are some of the complications that have arisen. UCL's Beacons for Public Engagement programme has chosen to measure public engagement efforts qualitatively within an interpretative framework for its evaluation purposes. Through case studies, interviews and document analysis, they determine how their public engagement activities are experienced by different stakeholders (UCL, 2016b). While public engagement and public outcome-oriented research is now firmly on the map, clearly more clarification is needed on how researchers are working with external partners, what they are looking to achieve and how effective these approach have been.

How are social media an essential part of public engagement?

Informal strategies for science communication on new media platforms have grown exponentially in recent years, and in particular Stilgoe et al. (2014: 9) note that 'social media have revealed an enthusiasm for uncontrolled engagement among those interested in science'. Public engagement efforts online, which are aimed at building dialogue with non-academic partners, are now common practices among researchers.

The wider shift encouraging more dialogue in universities has come alongside the emergence of new interactive approaches to communication, which we explored in section 2.1.2 on the convergence model. Notably, we have seen the widespread adoption and success of incorporating strategies that aim to solicit feedback, both online and offline, in public health research. Web-based platforms have proven themselves as useful in this space, though limitations have also been acknowledged. Contributing to blogs, websites and forums, and making sure your work is generally visible online, are great first steps and are certainly important, but true engagement often requires more effort and deliberation than dissemination to ensure a two-way exchange actually occurs.

Alongside platforms and networks that are facilitating connections and encouraging social exchange, certain formats, such as data visualisation and podcasts, can and are supporting researchers to engage more effectively with wider, non-academic audiences. Though there are many difficulties involved in productive two-way exchange, social and digital media as sites of open, networked content production offer novel ways for the public to be witness to, participant and even co-producer in research today.

2.2.6   Impact

What is it?

In his piece exploring the opportunities that social, open media present for digital humanities disciplines, John Maxwell (2015: 7) argues, 'Education, publishing and scholarship are all cultures of transformation: at their highest levels, they seek to have a transformative effect on the world'. The final element of the Research Lifecycle pertains to the transformative impact that academic research can have in society and the tangible outcomes that can be attained through the Research Lifecycle that extend beyond the research environment.

What does it look like?

Research work is now deeply embedded in a range of 'open' activity which encourages wider collaboration, readership, engagement and participation. As we have explored in the previous sections of the Research Lifecycle, given the rapid advances in digital technology and the emerging discourses in open digital scholarship, academic researchers are already exploring and integrating these social elements in their research practices. But what comes after the dissemination and engagement phases of the Research Lifecycle? What is this activity building towards? And how is this effort evaluated and recognised? All these questions fundamentally deal with the central role of impact in academic life.

As mentioned in the dissemination section and its overview of the academic publishing system, historically, publication quantity and the specific publishing outlet have played a significant role in the assessment of academic work. One effect of the importance that the publication venue has played in academic career structures is that institutions and individual academics recognise research outputs as ends in themselves.

And yet this output-oriented 'publish or perish' evaluation framework stands at odds with the 'transformative effect' that universities also have as their mission.

With mounting pressure from funders and wider trends in academia, there is now a growing emphasis on researchers, projects and institutions to demonstrate outcomes in society. The UK's Research Excellence Framework (REF) was the first national research assessment body to include social impact as a major part of its evaluation criteria: 20 per cent of the assessment criteria (and, thus, 20 per cent of REF-related research funding) dealt with whether the research in question could demonstrate 'an effect on, change or benefit to the economy, society, culture, public policy or services, health, the environment or quality of life, beyond academia' (HEFCE, 'REF impact', 2016).

Certainly, it is a good thing for researchers to consider the wider implications of the work they devote their working lives to, and many have welcomed the Impact Agenda as an opportunity to receive credit for research-related activities that did not previously fit within the strict confines of academic work (Smith et al., 2011). But there have also been a general backlash and scepticism over whether impact can be planned for and, indeed, if it can even be assessed. Dr Kathryn Oliver explores these concerns in a 2014 blog post for UCL titled, 'What's the impact of the research impact agenda?':

> The fundamental point is that we don't understand well how scientific evidence contributes to societal, political, physical, or economic outcomes – and acknowledging that makes a straw man of the direct impact metrics so many rail against. To me, this argues that we need better understanding of these processes, and a greater and more nuanced range of methods to understand what valuable and high-quality science looks like. (Oliver, 2014)

Simon Bastow, Jane Tinkler and Patrick Dunleavy have provided a comprehensive exploration of these processes in *The Impact of the Social Sciences: How Academics and their Research Make a Difference* (2014). From the extensive quantitative and qualitative analysis, it is clear that impact, as a complex social process more generally, but especially in the social sciences, is a very diffuse, non-linear and incremental process that occurs over time and involves a range of actors and influences. Figure 2.8 (which is from their book) does well to capture how scholarship (represented on the left, according to Boyer's model) filters through a range of interfaces towards its eventual impact.

The opening of the scientific process, as exemplified through wider access to electronic versions of research publications as well as researchers' engagement with non-academic partners, may be partly related to the cultural shift from deficit to dialogue discourses explored above, but is certainly also a reflection of the growing concern on behalf of the funding agencies to audit, record and assess the social and economic impact of academic research to provide the taxpayers with a return on research investment. Impact analysis frameworks are emerging across the Higher Education sector as universities are increasingly required 'to identify, analyse and articulate the actual and potential impact of their research' (NCCPE, 2013: 5).

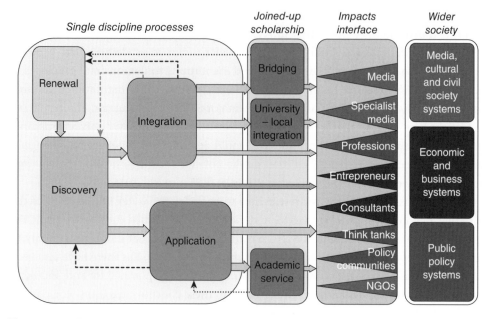

Figure 2.8   Social science impact interface

Credit: Bastow et al., 2014

A final aspect worth considering in relation to impact that complicates its measurement and role in society is that impact is not necessarily positive: academic research can be mistranslated, misapplied or generally misappropriated in a number of ways. Patrick Dunleavy (2014) recounts how the UK Treasury misapplied his 2010 research in their estimation of the costs of an independent Scottish Government in the run-up to the 2014 Scottish Referendum. As noted by the LSE Public Policy Group (2011) in the working paper, 'Maximising the impacts of your research: A handbook for social scientists': 'all societal changes create winners and losers and have unintended consequences, so that evaluating their net effects is always a non-trivial task.'

How are social media an essential part of impact?

Due in large part to the emergence of digital tools and technology that are 'disinter-mediating' the communication chain, research is filtering through to wider networks in business, civil society and government. Now that researchers and knowledge workers in universities, charities and governments are all responding to an explicit focus on audiences, reception, outcomes and accountability, advances and developments in media formats are receiving more prominent attention. Although it must be said that eventual impacts occur as the result of a series of factors and social and political processes, here is a brief look at how digital media are already proving to be an essential part of impact in the Research Lifecycle.

We have certainly seen a rise in 'Blogging for Impact' events in the UK in recent years, but there is still a great need for the academic community to develop a better understanding of how online outputs like blogs and other digital media are contributing to and influencing demonstrable changes in outcomes outside academia. For example, in Chapter 3, Dr Gabriel Siles-Brügge, an early-career researcher at the University of Manchester, has spoken about how his blog post on EU trade talks led to him giving evidence in a Select Committee. We look further across the chapters at how research-informed content has influenced a variety of policymaking processes in the UK, the EU and the US.

There are a number of different actors and steps shaping academic research's impact in society, but a common thread that we have seen throughout these examples is the mutually-reinforcing relationship between dissemination (shares), engagement (feedback) and eventual impact (long-term outcomes) in the Research Lifecycle.

2.3 CONCLUSION

Inspiration, collaboration, primary research, dissemination, public engagement and impact: this Research Lifecycle is the backdrop situating the needs of researchers, Higher Education professionals, policymakers and knowledge workers, and these needs set the scene for the many emerging opportunities for using tools and new media in research settings.

As we've seen in each section of the Lifecycle above, the modern research environment is not just conducive to the use of digital and social media, but *depends* on them to make research a reality. While we have seen traces of the social elements of research throughout history, this dependency is all the more apparent in the 21st century. As John Maxwell (2015: 5) writes, 'a natively digital scholarly work probably needs to be social in ways that traditional forms were not'. Traditional modes of research required particular media, but today's research requires a range of different media. Now that we have established a firm foundation for the different processes involved in research today, the rest of the book will explore the variety of available media being used in research settings and how to find the right ones for your research aims.

The following chapters will explore this social potential of research by looking at how academics are currently using blogs, podcasts, data visualisation, videos and photography. We have given a brief snapshot of how these formats are currently working across the Research Lifecycle, but the following chapters delve further into how to get started using these tools and how they can be employed effectively throughout the Research Lifecycle. We also provide case studies and refer back to our Lifecycle phases and diagram. We hope that these case studies will inspire further experimentation with social media and the social nature of research, while also recognising that social media are not merely a one-way conduit for research but are also being shaped by social processes themselves. Chapter 7 will return to the potential consequences this will have, both positive and negative, for how research is conducted and constructed.

2.4   FURTHER READING

Bastow, S., P. Dunleavy and J. Tinkler (2014) *The Impact of the Social Sciences: How Academics and their Research Make a Difference*. London: Sage.
Comprehensive background to the impact agenda, how research impact is achieved, measured and evaluated, and how these processes can be improved.

Goldman, A. and P.J. Vogt (n.d.) 'Reply all.' [Podcast.] *Writing on the Wall*. https://gimletmedia.com/episode/9-yik-yak [Accessed 27 April 2016].
An excellent podcast episode on the use of the anonymous social media app YikYak at Colgate University, where the anonymity brought out a particularly vicious strain of racism. The episode features excellent discussion of the role of technology in society and specifically of its use in higher education.

LSE Impact Blog eCollection: Open Access Perspectives in the Humanities and Social Sciences (2013) produced in collaboration with SAGE. Available at http://blogs.lse.ac.uk/impactof socialsciences/open-access-ecollection/http://blogs.lse.ac.uk/impactofsocialsciences/ open-access-ecollection
This collection of blog posts looks specifically at the opportunities and barriers of open access to scholarly literature in the humanities and social sciences. Contributors from a range of different countries reflect on how open access is being implemented and what can be done to provide greater access and accessibility of scholarly content.

Okerson, A. S. and J.J. O'Donnell (eds) (1995) *Scholarly Journals at the Crossroads: A Subversive Proposal for Electronic Publishing*. Washington, DC: Office of Scientific & Academic Publishing, Association of Research Libraries.
This collection includes Steven Harnad's 'Subversive Proposal' original email and the full listserv correspondence on academic publishing and the sharing of academic journal articles.

Weller, M. (2011) *The Digital Scholar: How Technology is Transforming Scholarly Practice*. London: Bloomsbury Academic. Available open access in full-text HTML for online reading: www.bloomsburycollections.com/book/the-digital-scholar-how-technology-is-transforming-scholarly-practice.

The Digital Scholar looks in greater detail at how academic practice is changing and from where these developments have emerged. Weller provides essential theoretical grounding for an engaged, academic practice.

Woodfield, K. (ed.) (2014) *Social Media in Social Research: Blogs on Blurring the Boundaries.* London: NatCen.
A collection of blog posts on how social science researchers are using social media to conduct research.

3
CREATING AND SHARING
BLOG POSTS

You write in order to change the world ... if you alter, even by a millimeter, the way people look at reality, then you can change it.

– Baldwin (quoted in Wong, 2015: 11)

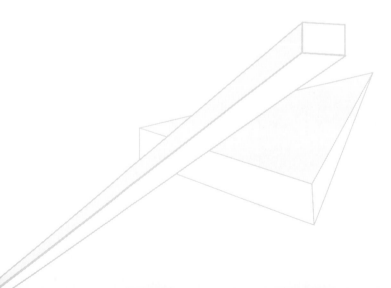

On 28 January 2015, Andrew Sullivan, journalist and editor of the politics and culture megablog *The Dish*, announced to the online world that he was quitting blogging, ending a run of highly popular commentary that began in 2000 and spanned several online platforms, including *Time* and *The Atlantic*. Weeks later, Sullivan revealed that he had parted company with blogging mainly for health reasons – the constant stress of leading a team which was responsible for publishing 40 pieces of content every day had finally taken its toll.

So with someone like Sullivan quitting this form of outreach, does this now mean that blogging is dead, as some have since commented? No, and far from it. The reality of blogging for the vast majority of those who use it as part of the Research Lifecycle is much more relaxed, enjoyable and rewarding. Creating accessible blog posts about your research, studies or projects can be painless and quick, and is a great route to opening up your content to wider audiences. To say that we are now seeing the end of blogging would be like asserting that we could see the end of books, or political comedy programmes; the form encompasses such a wide variety of styles and ways of doing that an end to it is just not possible.

In this chapter, we'll look at what blogging is, a bit about the history (and prehistory) of blogs and blogging, the different types of blog and blog post, and how to get started and keep going. We will also be looking at how best to share your blog's content using social media platforms, and how to deal with online comments and trolls. Whether you are an academic looking to tell your impact story, a communications professional seeking to promote your organisation's research, or a team leader looking for a way to share knowledge with your colleagues, this chapter will set you up with all the tools and knowledge needed to launch and maintain a successful blog for your project.

3.1 WHAT IS A BLOG? DEFINING BLOGS

Blogs were not always blogs. What we now know as blogs began life as 'weblogs' – literally writing on the World Wide Web – in late 1997 with the launch of John Barger's *Robot Wisdom WebLog*, where he presented lists of links that had caught his attention in the area of culture and technology (Rosenberg, 2010). The use of the term caught on, but, in 1999, Peter Merholz, writing on his website, *Peterme.com*, decided to split the word in two and discard the 'we' (Pole, 2010). Blogging, or at least the term, was born.

As is the case with social media itself, there are a number of definitions of blogs, which often depend on whether the author is a blogger or not. But the definition given by the Oxford Dictionary is helpful at this point, stating what a blog *is* without going into much detail as to what a blog is *for*: 'a regularly updated website or web page, typically one run by an individual or small group, that is written in an informal or conversational style' (Oxford University Press, 2016).

The first part of the definition is the important bit – regularly updated. Though blogs are a form of web page, the distinction which makes a blog different from a standard web page is an important one. One of the main features of a web page is that it is static; it is there to give you information that is more or less constant. The look, feel and even some of the content of the page might change, but the overall message will stay the same.

One incredibly important aspect of blogs which the OED's definition doesn't include is that they can also be a form of social network. Bruns and Jacobs (2006) emphasise the social nature of blogs, describing them as having a 'communicative intercast' which allows interaction, collaboration and co-creation.

Figure 3.1 Blog entry from Rebecca's Pocket blog, December 1999

Credit: Rebecca Blood (www.rebeccablood.com)

Now, back to the second part of that OED definition – written in an informal style. As Figures 3.1 and 3.2 illustrate, early blogs in the late 1990s and early 2000s were very much focused on informal reporting from individuals; commentary on people's social lives, reviews of music, products and travel destinations, for example. These sorts of blogs did tend to be written informally with much of the writing being done with an audience of friends or family in mind. While these types of blog do still exist, they no longer ultimately define the limits of blogging. It's worth keeping in mind that blogs came into being *before* search engines such as Google became widely used and effective; they filled a need for content filtering and curation that wasn't really available in any other meaningful way, at that time (Rosenberg, 2010).

If you take the definition of what a blog is at its most basic – a website with regularly updated content – then this encompasses a great deal of contemporary media. Most major online news sources are run in a blog format, with regularly updated articles and content. Editorials and comment pieces, as shown in Figure 3.3, which have survived and flourished in the print media's transition to the digital world, are, by definition, blogs. In discussing the style and format of blogging for researchers, the example of news editorials is a useful one to keep in mind.

Just as online editorials have different styles depending on the publication in question – witness the difference between editorials from *The Guardian*, *The Financial Times* and *The Economist* – there are different types of writing style researchers can choose from to present their ideas and content. Often these will depend on what is being written about – and when. We'll look at these later.

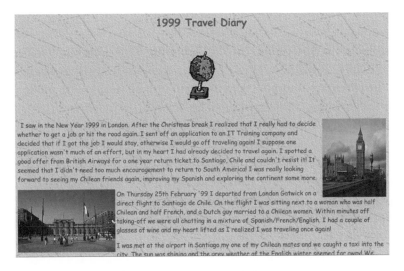

Figure 3.2 An early personal blog post: 1999 travel diary

Credit: Gregg McClurg

Figure 3.3 One example of an early comment blog piece

Credit: The Guardian Comment column from December, 2000

3.1.1 Differences between blogs and websites

There were just a handful of users, mostly based in academic research institutions such as CERN, when the World Wide Web began in the early 1990s. Since then, the web has grown to become home to nearly 1 billion websites (Internet Live Stats, 2016), with the vast majority of this growth occurring since 2010. By comparison, blogs numbered just over 300 million in 2013 (Pangburn, 2016). So with websites still vastly outnumbering blogs, what sets the two apart, and how does this inform how useful blogs are to academics, researchers and communications professionals? Table 3.1 explores the main differences.

As noted in Table 3.1, one of the key differences between a blog and a website is that it is regularly updated. A good example that shows this difference is a university's main web page. While its appearance will (hopefully) change from year to year as designers and developers take advantage of new software and faster internet connections, the information the page contains – course information, links to the library, events and university news – will remain constant. A blog differs in that it is updated regularly with new content. Regularly may mean once an hour, twice a day or seven times a week; what is key is that the content is refreshed systematically.

Web pages are also substantially different in terms of the roles that they play in the online community. A website is set up by a group, which can be a business (though these can be set up by a single person), charitable institution, church choir, etc., so as to give that group an online 'home'. For many, this home will describe what the organisation is, what its aims are, or perhaps it will be an online marketplace where people can go to buy products and services, such as the massively successful but often controversial Amazon.com. While a blog at its core is still a website, where the two part company is in their interactivity and how they reach out to those who read them.

Table 3.1 Characteristics of blogs versus websites

Blog	Website
Regularly updated	Static
Interactive, community building	One-way
Commentary and insight	Transactional information or news
More informal	Often formal
Very easy to publish new content	Dedicated CMS program needed, e.g. Dreamweaver
Easy to do at low cost	Quality = Cost

When thinking in the context of a university, a way to conceptualise this difference is to consider how an enquiries office is different from a public lecture. While students and visitors can walk in, call or email an enquiries office, the information that is provided about timetables, fees or course requirements is transactional and only one-way. This is the sort of information that would also be available on a university's website. Now, if we turn to a public lecture, where a speaker or group of speakers sets out a point of view in a 45-minute speech with discussion and questions from a panel chair, other speakers and the audience itself, we can see that there is sharing with a wider audience, members of whom then have a chance to respond with their own commentary and thoughts. The speaker then has the opportunity to answer or counter these new points, with the potential for them to be integrated into their own research or analysis in the future. This mirrors the structure of what is possible with a blog used for research, study or wider knowledge exchange. By its nature, a blog is a forum where writers are able to explain their ideas to a wider audience, and for the most part that audience is given an opportunity to respond to these ideas by adding comments, by contacting the author of the blog post or by discussion on social media platforms.

The next major point of distinction, as set out in Table 3.1, is in the flexibility in style. Since a website's major purpose is usually to impart transactional information to readers and users, its style tends to be formal and rigid. Blogs, on the other hand, are far less set in the type of information that they impart and in how they do it; their appeal to a wide variety of audiences can also mean a great range in style, both in terms of presentation and tone. For example, take the following two passages, both from economics blogs with wide readerships:

Example 1: Liberty Street Economics

The current policy debate is influenced by the possibility that the first-quarter GDP data were affected by 'residual seasonality.' That is, the statistical procedures used by the Bureau of Economic Analysis (BEA) did not fully smooth out seasonal variation in economic activity. If this is indeed the case, then the weak readings of the economy in the first quarter give an inaccurate picture of the state of the economy. In this post, we argue that unusually adverse winter weather, rather than imperfect seasonal adjustment by the BEA, was an important factor behind the weak first-quarter GDP data. (Groen and Russo, 2015)

Example 2: Mainly Macro

I have wasted far too much of my time killing zombies. This is what Paul Krugman calls ideas or alleged facts that, despite being shown to be wrong countless times, keep coming back to life. In terms of anti-Keynesian mythology, the zombie I have spent too much time on is that 2013 UK growth showed austerity works, but I've also done a bit on the mistaken idea that US growth in 2013 shows that Keynesian multipliers are zero. (I've been told that what I have done in the US case is deficient for a couple of reasons – neither of which I accept – but those saying this have never shown that doing it their way makes any difference. Instead they prefer to stick to gotcha economics. You can draw your own conclusions from that.) But these are particular episodes for particular countries – what about the big picture? (Wren-Lewis, 2015)

The *Liberty Street Economics* quote, from a group blog written by economists working for the New York Federal Reserve Bank, uses quite technical language and is aimed at other economists and policymakers, while the second is written in a much more personable and informal style, which has the aim of making the text more accessible to non-expert readers.

Not only do blogs allow much more freedom in terms of style and authorial voice, they also have much lower barriers to entry and publication compared to websites. While a blog can be set up within an hour for free, a website is much more complex. For the first near decade of the web's existence, what we now consider to be 'traditional' websites were the only option for those who wished to present their ideas and commentary to the wider online world. However, the late 1990s saw the beginning of free blogging services, largely commencing with the online diary platform LiveJournal, which was launched in April 1999, and was geared primarily towards more personal styles of blogging. Blogging as a platform became more established with Blogger (1999), Typepad (2003) and WordPress (2003). By 2013, WordPress had become the most popular of these sites, powering around 60 million sites worldwide (Tech2, 2016). All of these sites have had a free option for years, which allows users to create and style their own blog via their internet browser program; to begin, no more than an email address to register the blog is needed. Websites, by contrast, have historically needed a dedicated content management system (CMS) that required at least some knowledge of HyperText Markup Language (HTML) and PHP, a server scripting computer language.

The even relatively small skill level required in order to set up, build and maintain websites, including the more technical server-side requirements, has served as a significant barrier to entry to non-experts who wish to create their own websites, especially those working alone. Blogs, by contrast, use WYSIWYG ('What you see is what you get') text editors which are similar to widely-used text editing software such as Microsoft Word, and require no specialist knowledge. The ease of use of these blogging platforms is such that one could set up a blog in less than an hour, and update an existing blog in mere minutes. Ironically, WordPress is now used to power many sites that do not operate as blog sites (Wordpress.com, 2016), and instead would appear to be a more traditional website, such as CNN, Forbes and eBay.

3.1.2 Differences between blogs and journal articles

If you're reading this book as an academic or other knowledge worker, at this point, it is helpful to go into some of the ways in which blog posts are different from academic journal articles, as well as some of the ways in which they can complement one another (see Table 3.2). While we might argue that journal articles are one of the less productive ways that academics can engage with the public, they are still a 'necessary evil' in academia and research, as they are often key to how an academic's performance is assessed, with promotion and advancement closely linked as well. ·

Table 3.2 Major differences between academic journal articles and blog articles

	Journal article	Blog article
Length	8,000 words	800–1000 words
Frequency	Yearly	Weekly
Time to publication	Months/years	Days
Multimedia	Black and white charts	Colour, audio, video
Latitude for style	Nearly all examples have a formal verbose style	Broad ability to have own voice and to vary it
Yearly audience	Tens or hundreds	Potentially thousands
Availability to the general public	Paywall	Open access

The first major difference between journal articles and blog articles is in their length. The majority of journal articles are on the order of 8,000 words or so, while blog posts can be from 500 up to 1,500 words (they can be longer in some cases, but, for now, we'll concentrate on the shorter versions). Much of this length is a form of padding, whereby academic authors describe the previous literature and theory in the area as well as the methods used to arrive at their findings and conclusions. While the shorter length of blog posts means that there is less background information for readers, you can still provide links to background information – which can include your own journal articles – as hyperlinks in your text.

The second important difference is in the frequency of publication. The lengthy process of writing up, submission and peer review means that many academics are unable to get more than a handful of journal articles published every year. With a personal or multi-author blog, you can either control or have a large influence on how often you publish. A blog's much shorter length makes it easy to write much more often, be it once a week or once a month. Writing for blogs can give you more than ten times an academic's typical annual output of journal articles.

We will only touch on the issue of multimedia briefly here as that is something that we pick up in more detail in Chapters 4, 5 and 6. In terms of what they can illustrate, journal articles are still stuck in the 20th century – black and white images and figures predominate. Compared to what most people will encounter even in the pages of most newspapers, this puts journal articles even more firmly in the shade when it comes to communicating content. With blog articles, on the other hand, almost anything is possible, and more and more methods of illustrating content are becoming available.

Using colours and even interactive charts and graphs means that blog authors can show a great deal more information in the same amount of space – and in a way that is easier for readers to digest and to share, with social media buttons already installed on the blog page. Of a blogs' features, such as the ability to hyperlink to other resources, Putnam (2011: 4) states that they: 'move a blog beyond being a static block of written information and takes it into a faster paced environment, oriented toward rapid dissemination of information and hopefully progressing towards stronger collaborative knowledge'.

Another major point of difference between the two formats is audience size. Most journal articles, on average, are read by a small number of people (Eveleth, 2014), and citation does not guarantee that an article has been read (Simkin and Roychowdhury, 2002). Much of this is wound up with articles' lack of availability due to copyright restrictions. Blog posts, on the other hand, are open to be read by the hundreds, and often thousands; the latter is especially the case for multi-author blogs. If you want exposure for your work and commentary, blogging provides a much better opportunity compared to writing in a journal. Figure 3.4 shows the monthly readership for a successful blog post on one of the LSE's multi-author blogs over 2015.

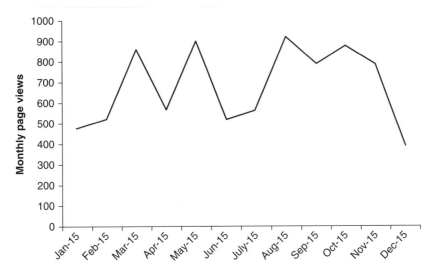

Figure 3.4 Monthly readership for a blog post

Source: LSE blogs data, the authors

As mentioned, the accessibility of blog articles is directly related to their lack of copyright restriction, i.e. they are not behind a sign-in paywall. By contrast, journal articles are closed to anyone outside an academic institution, unless the journal is open access. Despite the fact that academics are increasingly encouraged to prove the impact of their work on wider society, their main form of output is very often not available to those who could use it to make the greatest impact – policymakers working in government and NGOs.

It is worth noting that blog articles and journal articles are not mutually exclusive – indeed, they can support one another and help in their own creation. Ideally almost every journal article you write could also be written up as a blog article. On the other side, a blog article could be a useful way to present some early thoughts on a new dataset or test a new way of analysing those data. This could then be the kernel of a new article – or articles – that could be written up for publication. In an increasing number of cases, academics are contacted by paid blog editors, who ask them to send summaries of their articles for a multi-author blog or MAB.

So, to sum up, should you blog, it would need to be regularly updated; it would be a space for interaction and community; your blog posts would provide readers with insightful and understandable commentary; it would be easy to set up and publish; and could be done at a low cost.

If we have not convinced you yet, the next part of the chapter details some examples which might get ideas flowing on how blogging can be one of the most important things for your studies, research or project.

3.2 HISTORY OF BLOGS AND BLOGGING: FROM PAMPHLETS TO ORWELL - BLOGGING'S ANTECEDENTS

Putting aside the obvious technical and financial advantages of using blogs as a platform for the promulgation of your ideas and findings, the question, 'why blog?' remains. Digging down, the question is really 'why write at all?' As the quote from James Baldwin at the beginning of this chapter states, we write in order to change the world. Sometimes that process begins with the discussion sparked by a piece of writing which then leads to the sharing of ideas. It's the reason why Martin Luther wrote his *Ninety-five Theses*, which sparked the Protestant Reformation, why the Founding Fathers wrote documents such as *The Federalist Papers*, why Mary Wollstonecraft wrote *A Vindication of the Rights of Woman*, and why Karl Marx wrote the *Communist Manifesto* in 1848. As we'll see, today's blogs and commentary can help to alter the world by millimeters (in the words of Baldwin), and a good deal more.

While the advent of the World Wide Web now means that research and commentary can potentially be shared with an audience of millions almost instantaneously with geography representing no limit, it is useful to have a look at the history of sharing ideas before looking to the current landscape and examples. It's worth noting at this point that when we talk about sharing ideas, we're not just talking about academics. As mentioned in the previous chapter, anyone who works with ideas – what we refer to as 'knowledge workers' – will benefit from getting their ideas out there and into wider public view. You might be working in a health charity, think tank or an NGO and want to promote the findings of your latest research report. You might be a recently retired academic who still wants to contribute or share insights from many decades of research and investigation. Or you might be a young researcher taking a career break to focus on retraining, but still wanting to keep some skin in the game.

Writing dates back more than 5,000 years to ancient Egypt and Sumeria (Lerner, 2009), and has been the go-to conduit for sharing ideas across the centuries. The late 15th century saw the rise of the political sphere in England (Oliver, 2010: 14), centred mostly around London, which meant that there was an emerging market for the exchange of information and ideas.

Barlow (2007) argues that if there was to be a 'patron saint of the blogs', then it would be the politician, inventor (and Founding Father of the US) Benjamin Franklin. Franklin helped to lay the foundations through the postal service and the pre-Revolutionary War press for the public sphere of journalism, which enabled discourse and the exchange of ideas across time and distance in new ways. The nascent press encouraged by Franklin and those like him helped lead to important missives such as *The American Crisis*, published between 1776 and 1783 by Thomas Paine (Figure 3.5).

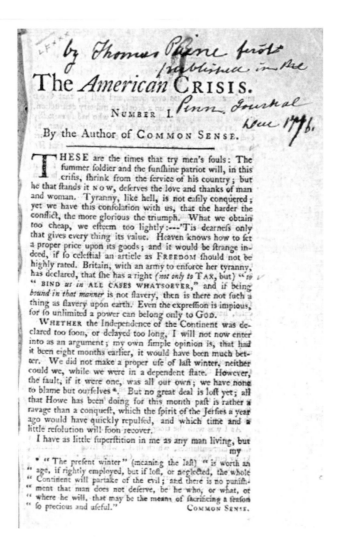

Figure 3.5 *The American Crisis*, pamphlet by Thomas Paine (2016/1776)

If we head across the Atlantic, pamphlets and papers were instrumental in another revolution of sorts in the mid-19th and early 20th century – the fight for women's suffrage. Pamphlets were used alongside articles in contemporary periodicals to agitate in favour of women's right to vote; these were often in response to, or met by, articles and pamphlets arguing that women should not be allowed to vote (Van Wingerden, 1999; Harrison, 2012). In some cases, pamphlets gave voices to those women who would otherwise not have been able to comment publicly in the society of the day.

Moving on to the 20th century, an important sharer of ideas was the writer and polemicist, Eric Blair – better known under the pen name, George Orwell. In 1946 Orwell wrote the now famous essay 'Why I Write' for the magazine, *Gangrel* (Figure 3.6). Two of the four reasons he gives as an explanation for his motivation to share his ideas are:

(iii) Historical impulse. Desire to see things as they are, to find out true facts and store them up for the use of posterity.

(iv) Political purpose. Using the word 'political' in the widest possible sense. Desire to push the world in a certain direction, to alter other peoples' idea of the kind of society that they should strive after. Once again, no book is genuinely free from political bias. The opinion that art should have nothing to do with politics is itself a political attitude. (Orwell, 1946: 5)

In many respects, Orwell's desire for his writing to have a political purpose – to move the world in a particular direction, as he states – is little different from the motivations of the pamphleteers of the previous five centuries. Orwell's magazine essays allowed him to publicly share his ideas on a regular basis with a wide audience and also receive feedback, both in person, and in writing, from his audience.

Is this sounding familiar?

Both Orwell and the pamphleteers that preceded him faced significant barriers to publication, be they in the form of an editor or the need to have access to a printing press. The accessibility of blogs means that the writers, researchers and students of today face virtually no barriers to publication, and can access a much wider audience than their predecessors could have ever imagined.

Figure 3.6   Cover Issue of *Gangrel*, containing George Orwell's 'Why I Write'
Credit: The Orwell Society

Barlow (2008) argues that blogs have managed to re-establish the public sphere in the early 21st century, acting as venues for debate in much the same way as the pamphleteers and essayists of previous decades did.

3.3 WHY BLOGGING IS USEFUL ACROSS THE RESEARCH LIFECYCLE

Reasons for knowledge workers to blog:

- Blogs can be inspiring
- Blogs can have policy impacts
- Blogs can influence the media
- Blogs can highlight your research

As we have seen, there's a huge amount of history behind those who want to have influence and change society and the world. Now we'll look at why blogging can help you to continue these traditions so that you are able to begin to make your own mark. Before we go into some more specific examples, here are a few reasons that bloggers have given over the years as to why they blog:

> I've been blogging now for a year and several months. I was inspired to create my own space in which to write by my desire to comment from a feminist perspective on contemporary theatre and performance, film, television, and novels in a free and unfettered space. *The Feminist Spectator* (Dolan, 2006)

> we hope that this blog will gain political science research greater attention and currency. ... Ultimately, we want contemporary discourse about politics centered as much as possible on the best kinds of evidence. *The Monkey Cage* (Sides, 2007)

> From the first few days of using the form, I was hooked. The simple experience of being able to directly broadcast my own words to readers was an exhilarating literary liberation. *Andrew Sullivan* (Sullivan, 2008)

> we might well ask whether it is wise to rest such power in the hands of a small group that appear democratically unaccountable. I'm not sure what the answer to this question is, but I think it is a question that should be asked more often. So, this is why I'm starting a blog ... I like to think that I have done enough in the past in contributing to the academic and policy debate to know when I have something interesting or important to say, or when I should stay quiet and do some more thinking. *Mainly Macro* (Wren-Lewis, 2011)

(Continued)

(Continued)

I enjoy it! At the risk of outing myself as a word nerd, I love writing about ideas and investigating social life, which is why I chose to become a sociologist. My blog gives me the opportunity to do this writing in a different way from the usual academic format. (Lupton, 2013)

blogging offers a space to write without censor, standard, and fear of 'what will the reviewers think!' ... In my personal blogging, I can talk about racism – and I often do. As a result, the words flow more easily. I do not stop after each sentence to agonize over what reviewer number 2 will say. And, this newfound ease in my writing extends into my academic writing, as well (even on 'perceived' race discrimination in my work on racist discrimination). *Conditionally Accepted* (Grollman, 2014)

3.3.1 Blogs can be inspiring

In the last chapter we acknowledged that ideas have to percolate in a researcher's mind for a time; they rarely emerge fully formed. We also looked at how science can function as a social network, with blogs helping to facilitate new ideas and new research. In a February 2016 blog post, a postgraduate student at Northwestern University wrote that she had been inspired by a blog piece on the *This Much We Know* blog to create a photo-based cartoon strip about Enterprise Social Networks, with the aim of informing colleagues of the benefits of using the tool for collaboration (Aminuddin, 2016).

Cammarata (2010) states in a blog post for the Smithsonian's National Air and Space Museum that a post written by a colleague on the institution's blog pushed her to join the conversation about how to ensure that children get the most out of their visit to the institution. She then ends her blog post by asking readers if they have any questions about bringing children to the National Air and Space Museum, another sign that blogs are very useful for public engagement.

Another example is Carnall's (2014) thoughts on whether or not the extinct species *Archaeopteryx* was a bird or not. Writing on UCL's *Museum and Collections* blog, Carnall was responding to a blog post on one of the blogs of the European Geosciences Union, *Green Tea and Velociraptors*, who posed the question of whether or not the ancient species was a bird.

3.3.2 Blogs can have policy impacts

As we have been discussing, blogs have the potential to achieve real changes in the world. One very good example is that of Dr Gabriel Siles-Brügge of the University of Manchester. In December 2013, Dr Siles-Brügge, along with a colleague, Ferdi De Ville, wrote a piece for the *Policy@Manchester* blog that detailed why, despite the arguments of its proponents, a recent round of EU–US trade negotiations would do little to significantly boost growth between the two (Siles-Brügge and De Ville, 2013). This blog garnered a great deal of media coverage and media appearances, and also led to Dr Siles-Brügge giving evidence to a UK House of Commons Business, Innovation and Skills Select Committee as part of its enquiry into the effects of the Transatlantic Trade and Investment Partnership on the UK economy (Linton, 2014; Waddington, 2015).

Heading across the Atlantic, a quick glance of the US Congressional records shows that blogs are frequently being used as evidence by legislators. For example, in March 2014, Senator John Barrasso of Wyoming mentioned a blog post written by Casey B. Mulligan, a professor of economics at the University of Chicago, in the *New York Times* (Mulligan, 2013). In the post, Mulligan wrote on the potential for increased demands on healthcare under the new Obamacare law (113 Cong. Rec, 2014a). In September 2014, Representative Keith Ellison of Minnesota cited a blog post from the Data Transparency Coalition which examined the ramifications of House Resolution 5405 on data transparency (113 Cong. Rec, 2014b). In July 2015, Democratic Senator Patrick Leahy of Vermont asked that a blog post by Professor Carl Tobias of the University of Richmond be read into the Congressional Record. The blog post, which originally appeared at *The Hill's Congress* blog, referred to the Senate's need to debate and vote on President Obama's nominees for the US Court of Federal Claims (114 Cong. Rec, 2015).

3.3.3 Blogs can influence the media

Not only are blogs capable of having direct impacts, they can also steer media narratives in certain directions as well. In 2012, Jack Monroe started her blog, *A Girl Called Jack*, which narrated her experiences as an unemployed single mother in Southend. Her blog became very popular through publishing recipes that would be able to feed families at low cost. Monroe rose to prominence through her blog, which led to guest columns in *The Guardian*. Through these columns Monroe was able to advocate for those on low incomes, initially through a series called *Austerity Bites* (Monroe, 2013). Monroe is a great example of someone who was able to use their blog as a springboard to comment on wider issues that affect society. In 2014 the UK parliament's All Party Parliamentary Group on Food Poverty heard evidence from Monroe on her experiences in using food banks. This evidence was later mentioned by MP Frank Field in a debate on the Modern Slavery Bill (House of Commons, 2014).

3.3.4 Blogs can highlight your research

Your ideas are important. If you're a student or an academic, you've studied for many years, gaining PhDs and other qualifications at great expense to get yourself into a position where you are able to investigate problems and issues that you are passionate about – or at least interested in. If you're a communications professional or a knowledge worker in charities, think tanks or NGOs, it is likely that you are responsible for sharing this research and information with the exact same aim as the former group. Whichever group you're in, blogging about your work can be the way to open it up to a wider audience, and we'll now look at some recent examples of blogging success, particularly with regards to disseminating research to academic and non-academic audiences.

For the time being, the major research output for academic research is the journal article or research paper. And while a great deal of time and effort go into producing these, there is growing concern at the readership (or the lack thereof) of these articles due, in part, to publishers' paywalls. While there seems to be limited evidence for the oft-mentioned statistic that 90 per cent of academic articles are never cited (Remler, 2014), it does seem that disagreements on this issue tend to circle around a number that many academics would wish was far smaller (Eveleth, 2014). Whatever the number, paywalls put up by academic publishers undoubtedly play a role in restricting readership and therefore citation counts.

Facing declining print revenues, recent years have seen media companies increasingly throw up digital paywalls for online journalism (Pickard and Williams, 2014). In academia, publishers' paywalls operate in much the same way, but at obviously much higher cost, essentially closing off content to those who are not based in an academic institution or cannot afford the cost of a journal subscription (we look in more detail at traditional forms of research dissemination in section 2.2.4 – the Dissemination section of the Research Lifecycle). According to Tillery (2012), the price of an annual subscription ranged from $46 per year for a history journal to more than $2,200 for chemistry. Bergstrom (2014) estimates that the median annual subscription for a journal in the social sciences is $606. Paywalls have been widely criticised by those inside (Taylor, 2012) and outside (Kendzior, 2013) the academic community. Arguments against include that paywalls often exclude those from reading who need it the most, such as journalists, policymakers and activists, and that since research is publicly funded, then it should be openly available (Taylor, 2013), rather than forcing readers to pay three times – once to fund it, again in subscription fees, and finally for its publication and dissemination (Brienza, 2014). The high price of accessing academic research has led to some recent efforts to 'free' it, including Sci-Hub, which has been scraping the internet for paywalled journals and making them available (Van der Sar, 2015; Resnick, 2016), and the #Icanhazpdf Twitter movement where users without access request that those who do have access to journal articles anonymously send them a PDF of the desired article (Mohdin, 2015).

For your research to have an impact on the world outside your doctrinal niche, it needs to be read by a wide audience, including those who may be able to use your work to make changes, such as policymakers and politicians. Without this wider audience, there is a real risk that your research will have little impact beyond academia.

In the past, publishers and authors had to work to understand their audiences; in the 18th century this was in the form of following up letters and missives sent in by readers on various issues (Barlow, 2007). Into the 19th and 20th centuries, newspapers, which are the most similar print form of what we know today as online journalism and blogging, could rely on circulation figures, but still had to do close analysis to understand who was reading and their reading habits (Lain, 1986).

Ong (1975) somewhat despairingly talks of audiences as 'fictional' in the minds of writers, with there often being little sense of who will actually be reading the written work when it is published, beyond who they imagine them to be. Even by the end of the 20th century, audiences were still seen as largely unknowable: 'The audience for most mass media is not usually observable, except in fragmentary or indirect ways' (McQuail, 1997: 2).

The idea of a 'mass' audience poses problems too. Most academics are specialists of one sort or another, who are seeking to inform a specific audience, which is interested in certain topics, rather than a mass public (though there will of course be times and topics that could and should be of very wide interest to readers).

While print media still exists, the old barriers to entry remain and, what's more, the explosion of free content online has undermined the importance and impact of what some have half-jokingly termed the 'dead-tree press' (Newhagen and Rafaeli, 1996; Bly, 2007). In the digital era, authors now have access to a plethora of metrics, and it is these metrics that illustrate the impact (another issue that has become increasingly important given the 'impact agenda' in the US, the UK and other Western countries) and importance that blogging can have for academics and researchers who wish to share their findings and commentary, as well as participating in debates about ideas (Kjellberg, 2010; Putnam, 2011).

It is also important to stress that the benefits of blogging can often go beyond the idea presented in the blog post itself; blogs can be the most important gateway to your research for those who you may feel need to read it the most. Blog posts which summarise research papers can then boost the readership of those papers themselves if they are linked to. In 2011, two World Bank Senior Economists, David McKenzie and Berk Ozler, set out to measure the effects of economics blogs on the dissemination of research papers. They found that when well-known economics blogs, such as *Freakonomics* and *Marginal Revolution*, mentioned certain research in their entries, this led to these papers getting up to three years' worth of hits in the space of a month. This massive spike in readership occurred despite the fact that only around 1–2 per cent of readers clicked through to read the abstract of a research paper (McKenzie and Ozler, 2014) (see Figure 3.7).

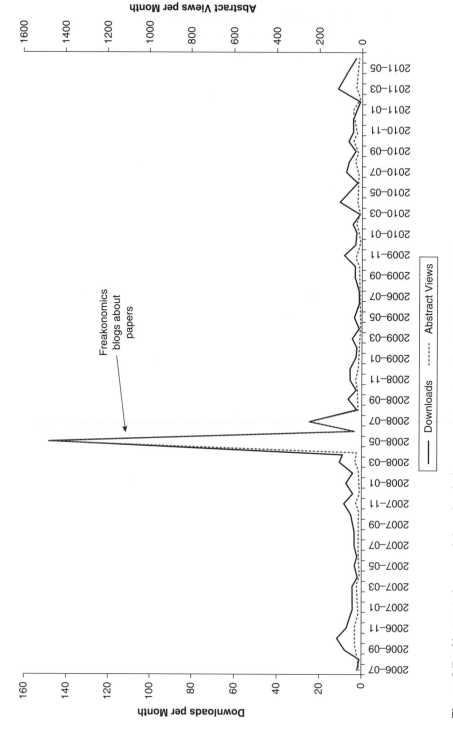

Figure 3.7 Abstract views and downloads for papers mentioned on Freakonomics blog

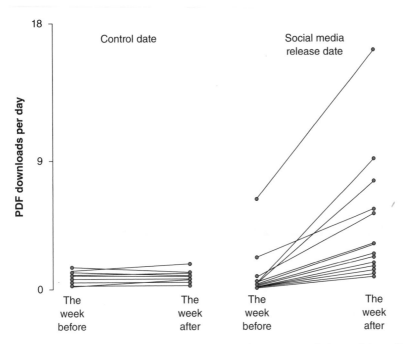

Figure 3.8 HTML downloads and PDF views of articles promoted via social media

Credit: Allen, H.G. et al., 2013. CC BY-SA

In a similar vein, Allen et al. (2013) found that social media releases – including via blogging – increased the number of HTML downloads and PDF views of articles on clinical pain, as shown in Figure 3.8.

Potentially, because of academic paywalls, blogs are increasingly becoming places where policymakers and others outside academia come for academic comment and analysis. In a recent study, Avey and Desch (2014) surveyed more than 230 US national security officials on how they get information from international relations scholars. They found that blogs were a significant source of information for policymakers, though classified US government reports were (unsurprisingly) still more important (see Figure 3.9).

It's important to note that in their study respondents noted that the lack of time to read longer reports was a big constraint. This is an area where blog posts can really shine. The short format that they entail is far more attractive than a 20-page policy report. Rounding out their assessment, Avey and Desch (2014: 244) state that:

> Our recommended model is one in which a scholar publishes his or her findings in traditional scholar outlets such as books or journals but also writes shorter and more accessible pieces reporting the same findings and telegraphing their policy implications in policy journals, opinion pieces, or even on blogs.

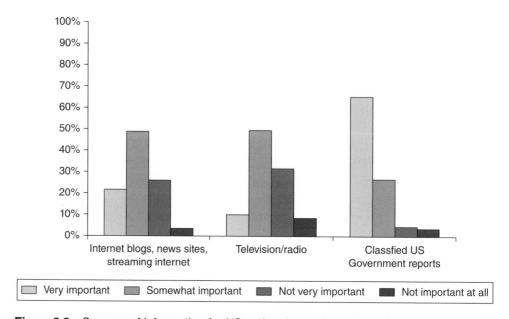

Figure 3.9 Sources of information for US national security policymakers

Credit: Avey, P.C. and Desch, M.C., 2014. Used with permission of authors and of Oxford University Press.

A similar study in the UK of 340 senior civil servants in government by Talbot (2014) found that research reports (79 per cent of respondents) and academic journal articles (55 per cent) were very important sources of academic research, but also that academic blogs were growing in importance. Of the civil servants surveyed, 14 per cent had used an individual academic blog, while 9 per cent had used a university-based academic blog. The study separated social media from blogging, but 27 per cent of academics had made use of Twitter, Facebook, LinkedIn or another similar social network as sources of academic research. It's likely that these sources will also incorporate links to academic blogs as well. What these two studies make clear is that policymakers are reading academic research, and not just in journals. Blogs are becoming increasingly important for civil servants and policymakers.

Academics and researchers have been blogging for the best part of a decade, but the practice has only really become mature and mainstream since around 2010. Since the late 2000s, newspapers, potentially as a response to declining print sales, have increasingly invested in or acquired online journalism and content, including blogs. In 2009, *The Washington Post* took on political blogger Ezra Klein to write *Wonkblog*, while a year later Nate Silver's *FiveThirtyEight* election forecasting blog was licensed by the *New York Times*. Both met with great success. By 2011, *Wonkblog* had become the most read blog (Narisetti, 2012) on the *Washington Post* site with millions of hits a month, and Silver had successfully predicted the outcome of the 2012 election in every single US state

(Griffel, 2012). By the end of 2014, however, both Klein and Silver had outgrown their respective stables, and left to found their own independent data journalism oriented outfits: *Vox* and a re-booted *FiveThirtyEight*. In 2013, the *Monkey Cage* multi-author political blog announced that it had been bought out by the *Washington Post*, a mutually beneficial deal for both parties. The *Post* gained a stable of already well-known political bloggers to add to its online offerings and the *Monkey Cage* bloggers were given access to a much larger audience than they previously had.

An outgrowth of this blurring of the line between blogging and journalism is that the pace of commentary has increased seemingly exponentially. While academic blogs and blog collectives produce content weekly or daily, the output of a *Washington Post* or a *Vox* has a frequency of hours – or even minutes. Because of these time pressures, many of these outlets have taken on roles as unofficial translators of research, summarising research papers and journal articles, and often asking the author for commentary as well.

While it is doubtless flattering to have one's research picked up in such a way by media outlets who measure their readership in the millions, this effectively robs the researcher of a voice when it comes to their own work. This can even cause problems between the online giants of data journalism. In April 2015, *Vox* was accused by Nate Silver of repurposing a chart from *FiveThirtyEight* without attribution or a link back (Arana, 2015). While this was fairly quickly resolved, with *Vox* editor Matthew Iglesias adding a link to the original post, this did highlight the risks that original content producers can face. It also proves that even if an author's work is hijacked in this fashion, contacting the offending party can often mean that the original work will be properly attributed.

Given that research papers often take years (Björk and Solomon, 2013) to pass through the peer review and publication system of an academic journal – and there is evidence to suggest that these times have been fairly static in recent years (Himmelstein, 2016) – researchers have ample time to prepare a summary blog article to coincide with the publication of their research. Publishing a blog summary ensures that the author is able to set out the most important findings of the report, and how important they are, and set the narrative around the research; those who conducted the research are attributed in the correct way, along with contact details, for further discussion; and journalists have an easy source for charts, graphics and quotations directly from the author when they do report on the research.

Another recent innovation is also worth noting here: *The Conversation*. Beginning in Australia in 2011, *The Conversation* employs professional journalists who work with academics to take their commentary to a wide audience. This content is then syndicated to online and print outlets across the world (Khan, 2016). *The Conversation*'s model shows the real potential of how blogging platforms can be used to increase the dissemination of research and commentary, by merging the world of journalism and academia into a multi-author blog (MAB) (see, for example, Figure 3.10).

Figure 3.10 *The Conversation* blogging platform

Credit: Simon Reich, Rutgers University Newark, writing for The Conversation

3.4 HOW TO CREATE AND MAINTAIN SUCCESSFUL BLOGS AND BLOG POSTS

Now that we've looked at what blogs can do and the kinds of impacts they can have, we'll turn to a more practical discussion of how you can write or create your own effective blog site. In this section we will look at the two main types of blog site, as well as some categories of blog post writing. We will also discuss how to make your own blogging better through choosing the appropriate style and tone for your audience.

3.4.1 Different types of blog and blog post

Single-author blogs

The first type of blog is the single-author blog, which we'll call 'SAB'. This is a blog which is set up, updated and promoted by one person. As previously discussed, this format is incredibly quick and easy to set up and update, but when considering going down this route, researchers and writers should be mindful of the advantages and disadvantages, as outlined in Table 3.3.

Despite the disadvantages, there are many examples of very successful SABs, many of which are embedded in the online news media. *The Conscience of a Liberal*, the blog of the economist Paul Krugman, is based on the *New York Times'* website. Simon Wren-Lewis' previously mentioned blog, *Mainly Macro*, on the other hand, is a relatively simple Blogspot-based enterprise.

Table 3.3 Advantages and disadvantages of a single-author blog (SAB)

Advantages	Disadvantages
• Author retains complete editorial control	• Risk of blog becoming too inward-looking, academically 'niche'
• Author has complete flexibility when content is posted	• Time pressure and need for constant level/interval of posting may be difficult given other commitments
• Author can tailor narrative and message of blog	• Heavy promotion required to overcome lack of interest at blog's inception

Moving on from economics-based blogs, Deborah Lupton maintains a blog called *This Sociological Life*, which encompasses her work on digital sociology, social media and critical digital health studies. Jill Dolan runs *The Feminist Spectator*, a blog which mixes commentary on the media, film, performance and the arts in general. Her blog is something of a hybrid, posting content which is academic along with personal reviews of the arts. Another blog which mixes academic commentary with more personal reflections is Pamela Gay's *Star Stryder* blog. Here, Gay blogs about astronomy, but also on the difficulties that women and early-career researchers can face in working in academia.

The single-author blog model fits researchers and knowledge workers who are committed to writing relatively often. This need for content at short intervals often leads to blogs which can have a mix of academic content and personal reflections, or commentary on the profession. Such blogs recognise that knowledge workers have more to say than just what they've found in their research, and as such can be a valuable outlet for personal expression.

Multi-author blogs

The second type of blog is the multi-author blog (or 'MAB') (Dunleavy and Gilson, 2012). This style of blog can take one of several forms:

- A collective based around one subject area that has shared editorial control (e.g. *Crooked Timber* – http://crookedtimber.org – is a blog collective based on around scholarship in international relations).
- A blog with a central managing editor, responsible for content commissioning, editing and posting (e.g. *LSE PPG* blogs).

The MAB is distinctly different from the SAB in both its form and operation, and ultimately provides those who may be new and possibly unsure about academic blogging with a good inroad to getting started. One of the most basic advantages that the MAB has is that the workload is shared across a number of authors, which can number anywhere from two or three to the hundreds. This takes the pressure off any one author to constantly

Table 3.4 Advantages and disadvantages of multi-author blogs (MABs)

Advantages	Disadvantages
• Posts are edited by MAB blog editor to ensure impact is maximised	• Authors lose editorial control – may disagree on best way to maximise impact
• Dissemination (including via social media) undertaken centrally by blog editor	• Authors lose opportunity to gain reputation from their own platforms
• Blog article benefits from being in 'stable' with others – dissemination can snowball	• Authors can lose 'voice' if blog is one of many being published

produce content, while still allowing content to be published on a regular (often daily) basis. Table 3.4 outlines some of the advantages and disadvantages of MABs.

Just as there are different types of blog, there are different types of blog post – even more so, in fact. Whether or not you use a certain type of post will depend on the message that you want to convey. Depending on your blog, you may make use of one type every week or even every day, while some may not suit your style of writing or topic and thus are never used. They will also depend on the type of blog you're writing for and on your audience – some styles will be more apt for MABs, others for an SAB blog.

Before we consider the various types of blog post you can choose from, it is worth considering how much to write and what will appeal most to your target audiences. You might have thousands of words to write about the mechanisms used by cells to repair DNA or ideas on how to create the jobs vital for economic revival in the EU, but will your audience really be able to use or find your work if it is in an unsuitable format? Whatever the topic, a blog post should be a minimum of 500 words and a maximum of 1,500 words. This length gives you enough space to introduce your topic and explain your thinking. Some types of blog post will lend themselves to different lengths. Research updates, for example, will tend to be shorter, while commentary pieces will tend to be longer. But remember, audiences are impatient. Content on the web is not like a book or a magazine that can command readers' (near) undivided attention. Evidence suggests that the vast majority of readers do not read any web content – websites or blogs – to the end (Manjoo, 2013). So remember, the shorter, the better.

What to write about: Types of blog post

'Write about what you know' is an old chestnut, and for good reason. It is a piece of advice that can be as equally applied to writing for your blog as for any other medium, and leads us to the first category of blog post: **writing about your research**. Posts which outline the background and aims of your research are often an ideal post to start with when beginning your own blog. This will introduce readers to you and your work and will be a good signal which will show them if your blog will be of interest to them as it continues and progresses.

This then leads to the next type of blog post: **research updates**. A major problem that early-career researchers or those new to the idea of blogging often complain of is that they have nothing to write about. But they will almost always be working on *something*. Why not blog about it as well? For a research update blog post, you don't have to wait until you've completed fieldwork or have some findings to report – though those are all worthy topics. An update could just as easily be about your latest thoughts on how to tackle your research question, reflections and comments on papers and research reports you've recently read, or reactions to conversations you've had with colleagues and peers about a project. In many respects, this type of post serves almost as a form of research diary – but an interactive one.

As you develop and grow your skills as a researcher, **you will learn things – and you should share these with others**. Whether it's how best to take advantage of the newest version of SPSS or a thinking exercise that worked for you, these are all valuable insights in which others will be interested, and that will in turn draw people to your blog. Gone are the days when early-career researchers had to attend a departmental seminar to learn about new research techniques or areas of study – this sort of information is now only a Google search away. By outlining in an easy-to-understand way what you have learned, as well as the problems which led to your need to understand something new, you can help others and give them a reason to come to your blog, and to return to it. 'How to' guides are gaining in popularity online, and ones that are written by academics for academics (or students or knowledge workers) will always be popular among those global communities (Figure 3.11).

Another type of blog post that you are very likely to have seen already is the **commentary blog post**. These can take two forms: the commentary post based on current events, but not linked to your research (good), and commentary with a direct 'hook' linking to your research in some way (better). The first type is very common on commentary sites such as *The Conversation* and *The Guardian*. They are typically commissioned by a central editor who wishes to publish commentary on a current event. This is also the case for many MABs as well. These posts typically introduce the topic that is in the news and then give some degree of insight into what's been going on, and why. While this commentary will draw on a researcher's in-depth knowledge of a topic, it will tend not to include evidence such as graphs, charts or statistics. These types of post are very useful for getting your name out as a commentator on topics that are of interest to the media, but are less so in terms of getting your own research out.

Posts which link back to your own research are preferable for academic researchers, although, somewhat predictably, they are harder to do regularly. However, unless you are investigating something truly obscure, your research in all likelihood will have some relevance to a topic in the public consciousness at some point. This is why it is important to keep abreast of national and international news and debate as much as you can – you never know when your opportunity to contribute might occur. Given the constraints of

Figure 3.11 *LSE Impact Blog* 'how to' guide on academic presentations

the media cycle, your research does not even have to be '100 per cent fresh'; if it is fairly recent, it will still be relevant. For example, in the summer of 2015 the UK was gripped by debate over the crisis of refugees in Calais attempting to come into the country, and the US presidential election campaign was heating up over comments from billionaire Donald Trump on immigrants from Mexico. In another example, in 2016 the UK saw junior doctors strike over a disagreement about working on weekends. These debates would be ideal entry points for researchers looking into (but certainly not limited to) any one of the topics listed in Table 3.5.

By establishing yourself as a commentator on current events by using your own research as a hook, this not only provides content for your own blog, but also allows you to showcase your writing for media outlets, which can enhance your reputation as an expert commentator and raise the wider profile of your organisation.

The last type of blog post is the distant cousin of a research report: **a report on a conference** or other event. Early-career and long-established researchers alike attend a number of conferences, seminars and similar events every year. Why not use these as an opportunity for comment? How often have we gone to such events, taken notes and then never thought about them again afterwards? Writing reports about these types of experience means that this process of reflection becomes a built-in part of your research process and experience. Many speakers at events and conferences will also have their own presence on social media and blogs of their own. Writing a reflective or critical blog post about the event means that the discussion does not end when the live Q&A does. It also gives you an opportunity to connect directly with the speaker and provides a way for others to contribute, even if they were not able to attend (see Figure 3.12).

Table 3.5 Example entry points for blog commentary

Role of immigrants in society

• The role of the media, religion, class, etc. in attitudes towards migrants • Theories of national identity vs 'the other' • Debates over the integration of migrants in society	• What do migrants want? • What makes certain countries so attractive to migrant groups? • Government efficiency and bureaucracy • Concepts of 'fairness' between countries

NHS and healthcare in the UK

• Training and recruitment for junior doctors • Provision of NHS services by private companies • Allocation of health services to communities in need	• Role and importance of the UK's National Health Service in society • Protections for NHS employees that stem from the UK's membership of the European Union • Inequality in health services cutbacks

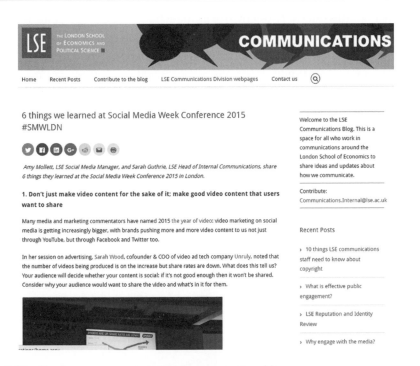

Figure 3.12 Conference report on *LSE Communications* blog

Credit: Conference report on LSE Communications blog. From http://blogs.lse.ac.uk/communications/2015/09/23/6-things-we-learned-at-social-media-week-conference-2015-smwldn/. Used with permission.

3.4.2 Tools for making successful blogs

Thinking about your blogging style and tone

While we discussed considerations of blog post length above, writing an effective blog post is about much more than that. Elements to consider are style, tone and evidence, all the while keeping in mind the needs of your audience.

By **style**, we mean how your blog post looks, from font to paragraph structures, to how you use images and infographics. On a very basic level, your blog post should obey the rules of good writing: lead with the best, as journalists do. You'll notice that – and may even be guilty of this – journal articles lead readers on a rather tortuous and lengthy journey through an Introduction, Literature Review, Methods and then finally on to the findings of the research after about 6,000 words or so. You need to throw out this way of thinking when it comes to writing blog articles (Dunleavy, 2014). An effective blog piece should begin with a short introductory paragraph – akin to a condensed abstract – which outlines the main message of what you want the reader to take away from your text. Given that the average online reader now only spends 15 seconds or less (Haile, 2014) on a website, this introduction may be your only way to catch a reader's attention.

Following from this one- or two-sentence introduction, give a brief outline of the topic, be it background to a current event you're commenting on, a problem or question you set out to solve, or a few details about an event or conference you attended. Then give the main take away: your findings, conclusion or general message to your readers. These should be in paragraphs that are relatively short, punchy and easy to read for a variety of authors. Figure 3.13 gives an idea of what your blog post as a whole might look like.

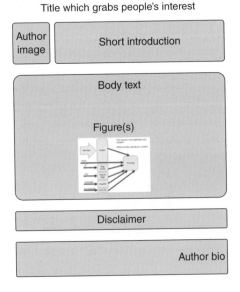

Figure 3.13 Template for blog articles

Subheadings are helpful in posts which are longer – around 1,500 words or so. Articles of this length should have at least two narrative-style subheadings which summarise the points made by the upcoming paragraphs. This will help guide readers through your text by letting them know what's to come.

When thinking about writing style, it is also very important to consider who you are writing for, and where they are. For example, you may be British, but primarily writing on American politics or history. While writing from the UK, you are much more likely to be well received by your target audience if you write with American spellings and idioms: fall rather than autumn, program rather than programme, etc. Earlier in this chapter we mentioned that the blogs at *The Guardian*, *The Economist* and the *Financial Times* had differing writing styles. *The Guardian* tends to be targeted at a more generalist UK audience, with *The Economist* and the *Financial Times* written with a more international, or even US-centric, audience in mind. The latter two publications also tend to cover economic and financial topics in a more in-depth fashion, so will use related terms much more frequently.

You may even wish to write your own style guide and pin it to your wall or desk, or make a note on your phone or tablet to remind yourself of how you want to write. A style guide would cover elements such as what type of spelling to use (US or UK), the minimum and maximum length of blog posts, how other blogs and academic articles are referenced and linked to in the text, and (possibly most importantly) who your audience is.

Titles and subheadings are incredibly important contributors towards making a blog post effective. They are your readers' signposts that let them know what to expect, especially if you're writing as an expert for others who may not be. Narrative titles and subheadings are key here. With blog readers often short on time and attention, short summaries are good ways to help make sure the main messages of your piece get across to them. For titles, you should avoid simply describing what you did, or the subject of your research, or teasing people by referring non-specifically to what you've found out. For example, rather than the unillustrative 'The effects of xxx on yyy', or 'Does xxx lead to yyy?', a better title would be 'How xxx means that yyy leads to zzz'. Such a narrative title tells the story of your blog post in a small number of words (fewer than 15 is best, if possible); this title is then much easier to pick up and be shared on readers' social media feeds. Examples include:

- How academia resembles a drug gang
- Chicago is far from being the murder capital of the US
- Profit from crisis: Why capitalists do not want recovery, and what that means for America
- Should the UK stay or go? The economic consequences of Britain leaving the EU
- Spain has reason to be concerned at its latest unemployment figures
- There was no rise in Scottish nationalism: Understanding the SNP victory

An alternative title formation, which is becoming increasingly popular with multi-author blog sites, acknowledges readers' aforementioned short attention spans and therefore uses very short titles, which are often in the form of a question:

- Why you should care about the local elections
- Is social media making people depressed?
- Doctors look after our mental health but who looks after theirs?
- This is how language might be mapped in your brain
- The most unnerving stories you can't miss this week
- Canada still doesn't know how much money it loses to tax havens

Tone is less prescriptive but just as important as the style of your blog. Most academics write in a relatively turgid and often exclusive (Lillis, 1997) or vague (Myers, 1996) style that can be very difficult for anyone outside their own discipline – let alone members of the public – to understand (Elbow, 2014; Rothman, 2014). Examples of this include the heavy use of Latin phrases, obscure academic and disciplinary terms, and extensive use of the passive voice. If you have made the decision to write a blog post, then you have also made a commitment to write for a wide audience, most of whom will not be academics. By writing in an academic style in a blog format, your blog's chances of success will be massively reduced from the beginning.

Even with this advice to not write 'academically', blog authors still have options of different writing voices. Some choose to go down a very conversational route, putting themselves at the centre. This technique often uses storytelling as a device to bring readers into the author's own research journey. Blog posts in this style often have the feel of a conversation in a social setting. While this form is great for communicating ideas to a wide audience, authors may be wary of jumping straight into this style if they have not previously written for a wider audience, and are used to a more 'academic' style of writing. In this case, a slightly more formal approach might be a good idea as it will be less of a jump from what you've written previously. Figure 3.14 shows some well-known blogs, grading them on a sliding scale from formal to informal styles of writing and content.

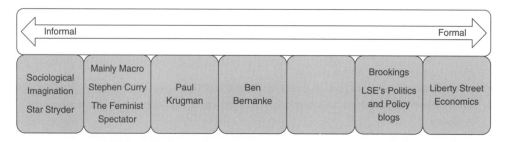

Figure 3.14　Range of blogs' writing styles from formal to informal

Evidence

One of the most important features of a blog post written by academics and knowledge workers – especially when it comes to commentary pieces – is that it should be evidence-based or present a logical argument or discussion that is defensible with evidence. Academics and students spend their time investigating the world with tools that their previous training, study and research have equipped them with, and this means that when you do find something out, or come to a conclusion, then this is based on something that you can point to. The contemporary media world – both online and in print – is strewn with opinion and commentary pieces that are based on the writer's 'gut feeling' with little other than emotion and an appeal to 'common sense' to justify the line of reasoning or argument that is used. As an academic or expert commentator, you can do better, by balancing the need to show your readers your work while not simultaneously boring them to death.

Using evidence to back up your commentary is not simply a matter of adding in a regression table from your latest article or even a graph. As discussed above, it is important to present your facts and arguments with your audience in mind. While equations and details of multivariate regression tests may seem like a good idea if you are presenting a suitably complex area of research, they will only turn off non-expert readers, something that could be harmful for a newly minted blogger who is just starting out.

In the majority of academic publications there is a need – often a requirement – for authors to 'show their workings', and include a review of previous scholarship on an issue and to map out in close detail the methods used to find data and reach conclusions. While blogging, writers must essentially cast these ways of writing aside. There isn't enough space (or reader interest!) in an academic's blog post for this kind of verbiage, and it would actively detract from the point being made if it were to be included. The key thing here is to be concise in how you use evidence and get your point across as early as possible. This is a technique that the LSE's blog editors have used and encouraged; many of the blog articles posted on LSE blogs use this format (see Figure 3.15).

Questions to ask yourself before you blog

So, you've decided that you want to start a blog about your research, studies or project, or you want to join with colleagues or others in your field of expertise in a multi-author blog. What do you do first? What are the essential steps to take for maximising success? In this section we will talk about the nuts and bolts of blogging on your own first, and then look at some considerations for participating in MABs.

As with any endeavour that is worth undertaking, starting to blog requires planning – or it does if it's going to be a success. Here are some questions that you should be asking before you get started, which will help you to set your expectations about how much work will go into your blog and what you will get out of it:

- What platform will you use for your blog?
- Who are your target audiences? Are they mostly academic or are they policymakers, students or the general public?
- What will your blog look like?
- What will it be called?
- How often will it be updated and how will you ensure it is updated regularly?
- What types of post will you have?
- How will you promote your content?
- How will you engage with your readers?
- How will you build networks to share your blog's content?
- Will you be using your research to help you to build a platform of advocacy with an end goal in mind?

Authoritarianism, not social class, is the dividing line between supporting and opposing Donald Trump.

Share this:

  In the nine months since New York billionaire Donald Trump launched his presidential campaign, many pundits and commentators have attributed his snowballing success to his popularity with white working class voters who also lacked a college education. Using new survey data, Jonathan Weiler and Matthew MacWilliams find that this characterization of Trump voter isn't accurate; rather than class or education, authoritarianism is the biggest driver of support for Trump.

With New York's primary results now counted and the probability of a general election contest between Donald Trump and Hillary Clinton increasing, it is time to reexamine the simplistic argument that, driven by economic populism, white working class Americans are a monolithic voting bloc who support of Donald Trump. This stereotyping of lower income and education whites as social and cultural conservatives ripe for Trump's rhetoric fits the conventional media meme, but is too simplistic.

Class is not the dividing line between Trump support and opposition. Authoritarianism is.

How do we know? We examined voters' feelings toward Trump by their education, income, *and* authoritarianism and found that it is authoritarianism, not class that appears to explain their views.

Our findings come from a national survey of 1800 voters. Authoritarian-minded voters were identified based on their answers to four parenting questions that tap the degree to which people prioritize order and social hierarchy. Voters' views of Trump were assessed using a standard feeling thermometer in which a zero represents a very cold feeling and a 100 a very warm or favorable feeling toward him.

Figure 3.15 *LSE USAPP* blog post showing 'front-loading' blog post's argument

Credit: http://blogs.lse.ac.uk/usappblog/2016/04/21/authoritarianism-not-social-class-is-the-dividing-line-between-supporting-and-opposing-donald-trump/. Used with permission.

We offer the same advice for **blog names** as we do in most other aspects of blogging – keep it simple. The most successful blog brands tend to have very short names: *The Monkey Cage, The Chronicle of Philanthropy, The NonProfit Times, Getting Attention, 1% for the Planet, Work in Progress, The Mischiefs of Faction, Crooked Timber, Mainly Macro,* to name a few. As these examples show, many blogs have achieved success without having too technical or descriptive a name. In these cases, the content that these blogs regularly post is what adds to their reputation, and the names are more of an easy-to-remember marker.

When deciding on your own name, first do your homework: what are other blogs writing on the same or similar topics called? At the very least you need to ensure that you do not use the name of an already popular blog. It is also a good idea to test out any ideas you have for a blog name on colleagues and friends, to see if it is something that they would consider memorable or interesting.

While there are a number of **blogging platforms** available, we really only recommend three, as they are the easiest to use and have the widest range of support for users: WordPress, Blogger and Medium. You might want to ask colleagues and friends what they recommend, or get stuck in and start exploring the homepage of each and choose whichever appeals the most.

> Many blog writers choose to write their blog posts out in Word or some other more dedicated writing software prior to adding them into a web-based platform such as those below. Most people will be familiar with Word and its competitors, and so it is often a better way to write blog posts without getting bogged down with formatting issues during the writing process. Once your text is finished, it's a relatively simple matter of pasting it into the blog platform in question.

First launched in 2003, **WordPress** is one of the most popular blogging services on the web, and is also used as a Content Management System for tens of millions of websites. As of late 2016, WordPress blogs are divided into two types – wordpress.*com* and wordpress.*org*. The latter is for more technically savvy users, and it is unlikely that you will need to worry about it at this early stage of your blogging career. Wordpress.com, however, allows you to very quickly set up a blog site, under its own domain, so your blog's address will be in the format 'https://example.wordpress.com'. LSE's blog platform uses WordPress, and it's a system that blog editors there have resoundingly found easy to use, compared to content management systems. This resonates with others such as Zhang and Olfman (2010) and Boulay (2013), who have used WordPress to design online materials and as a classroom blogging project.

Blogger is similar in function to WordPress. Launched in 1999 and taken over by Google in 2003, Blogger numbers its users in the hundreds of millions. The service also uses the subdomain blogspot.com, so your blog's address will be http://example.blogspot.co.uk.

The strength of these blogging platforms is that they are relatively easy to use for those who have minimal IT experience beyond Microsoft Word, and that they are customisable and extendable for those who do have more experience in this area or who want to learn.

Compared to Blogger and WordPress, **Medium** is very much the new kid on the blogging block. Begun in 2012 by two of Twitter's co-founders, Medium is even simpler than its blogging platform stablemates. Its interface is incredibly simple, placing the emphasis on writing – though images are supported. Medium is better suited for more occasional blogging as it essentially works as a massive multi-author blog, given the format is consistent across all those who are writing on the site. Medium is also designed more for writing blog articles directly on the site, rather than importing them from elsewhere.

Whichever blog platform you choose, you will have the option of applying a specific theme. Themes dictate the personality of your blog in terms of colour palettes, general design and functionality. While there isn't enough space to go into a great deal of detail about which designs to choose (and our recommendations would end up outdated within a very short amount of time), it is worth spending some time thinking about this. If your blog is more of a research diary, then you might want to use a blog theme that can showcase photos from your research. Alternatively, if you will be showcasing your writing, a theme which is less 'busy' may be preferable to one that has menus and widgets which may distract readers from the main text.

Today, every theme will recognise that readers will be accessing your work in a number of different ways, and so will adjust to make sure your content looks great whether it's being viewed on a PC, tablet or smartphone. Once your blog is up and running, take time to view it on multiple devices, different sized screens with different resolutions, or in different web browsers. It's important to remember that *your blog does not look like you think it does*. It's inevitable that you will be updating and reading your own blog on one or two different screens, so you will become used to what your blog is 'supposed' to look like. It is unlikely to look like this to the majority of your audience. Take time to study your blog's analytics to find out how people are seeing your work, and make sure you keep that in mind when designing and updating your blog.

For academics, students and researchers who prefer to join an existing blog collective or a larger established MAB, the process is fairly straightforward, with just a few considerations to keep in mind. The first is to make sure you've found a multi-author blog that best suits your writing style and research. Most disciplines have at least several MAB collectives, so you'll need to do some reading to find the best fit. Second, at this point it is helpful to have produced some writing in a blog format. The managing editors will probably be interested in reading something you've already written for a public audience or general reader. Third, you should also have a sense at this point of how often you'd like to contribute.

You may wish to send them only a one-off commentary post at this point, or you might prefer the option of sending them something once a month or so.

When you've found a MAB you'd like to contribute to, you'll then need to approach them with your offering or a pitch. Most MAB groups will have a dedicated group email address or editor – though this position may be rotating. Once you get in contact with your chosen MAB collective, they will then let you know about whether or not they are accepting new contributions and will send you information about their own style and editorial guidelines, and the timings involved for posting your blog article if and when they accept it. From the point of view of the managing editor of a MAB, it's often best not to send the full article first. Instead, send a short paragraph outlining your topic and any 'hooks' tying it to current events. This makes it easier for an editor to determine whether or not the piece is apt to run on their blog, or if it might be better placed elsewhere.

Once your article is published, the MAB's editors will also promote your piece on their social media feeds and any email lists that they maintain. This is a definite advantage that MABs have over single-author blogs; one post can reach a much wider and more diverse audience. For example, in 2015, an article on the LSE's *British Politics and Policy* blog, 'Youth unemployment produces multiple scarring effects', by Ronald McQuaid (2014), was viewed more than 7,000 times by people in more than 40 countries, speaking 15 languages. This was helped by the blog's over 40,000 Twitter followers and over 3,000 Facebook 'Likes', which were able to promote the post. On occasion, very timely blog posts can go 'viral'. In the summer of 2016, an article by LSE Professor Nicholas Barr on the UK's EU referendum had more than 500,000 visits in the space of a few weeks (it also crashed the LSE's blog servers!) (see Figure 3.16).

Figure 3.16   *LSE BrexitVote Blog* by Professor Nicholas Barr which went viral in the summer of 2016

Credit: From http://blogs.lse.ac.uk/brexit/2016/05/27/dear-friends-this-is-why-i-will-vote-remain-in-the-referendum/. Used with permission.

The importance of posting regularly

No matter which route you go down in terms of the way that you publish your blog articles, it is important to do this regularly. If readers know that they can expect new content from you once a day, or every two days, or once a week, or even once a month, then they are much more likely to continue to come back and read your material. The web is littered with the corpses of tens of thousands of blogs which were begun with good intentions, updated weekly for a month or two, and then went silent. It's worth remembering that even popular blogs can suffer high bounce rates (the number of visitors who 'bounce' off of a site after arriving and do not click on other links); the LSE's *British Politics and Policy* blog had a rate of nearly 86 per cent in 2015, with less than 32 per cent of visitors being returning ones.

As updating a blog regularly can be difficult, it's important to think about how often you can realistically guarantee writing. For most new bloggers, it is unrealistic to plan to post something every day, and even once a week can be a lot to aim for. It is much better to begin with regular but spaced-out postings, and then to increase your frequency as your blogging experience grows and if you have more to write about as your research or project progresses.

It's also worth considering implementing an editorial 'grid' system, which can allow bloggers to plan posts out weeks or even months in advance (see Table 3.6). When creating the grid, bloggers can populate it with anticipated events that might be springboards for blog posts. These can include conferences, consultations, or the anticipated release of new research in an area of interest. Aside from being a good tool for horizon scanning, an editorial calendar of this sort means that a blogger can have a better sense of the flow

Table 3.6 Sample editorial grid for a blog

Week beginning	External	Publications	Blog activity
1/8/16	End user conference – University of Sussex	–	Conference summary
8/8/16	–	Journal article live	Blog post summarising article
15/8/16	–	–	Research update
22/8/16	Government select committee report released	–	Post commenting on report + process
29/8/16	–	–	Guest post from colleague
5/9/16	–	–	'How to guide' on using podcasting software

of posts, and whether or not a certain topic or type of post is appearing too often, or not often enough. It can also act as a record of past posts, which can be a useful resource in developing new ones, as bloggers can reflect on what's gone before and what has been learned since then.

Writing and workflows

One of the biggest issues new bloggers face is the question of what to write and when to find time to write it. In terms of what to write, if you refer back to the section on 'Types of blog post', there are over half a dozen options for creating content, such as reporting from a conference, writing about a new skill or tip you've picked up, or a commentary piece. Whatever stage of the Research Lifecycle you find yourself at, you should be able to write a blog post based on one of these types.

Another issue that early-career academics and busy research assistants face is a lack of time, and this can be even more of an issue than being unsure of what to write about. While we can't magically create more hours in the day for you to write in, we can offer some tips on how to use the often very limited time that you do have to write blog pieces. The first part is preparation. One tactic is to create a simple blog calendar, marking out the various parts of your academic life and work that could lend themselves to a blog post. For example, if you know that you will be attending a conference in the third week of the coming month, you can write a note in your calendar to write a blog post summarising either your presentation or your reflections on the conference as a whole, or both. Publishing just a few hundred words and a photo or perhaps some of your slides will be a great marker of the event and a small but rewarding achievement.

The second way to help along your blog writing process is to note down your thoughts and observations. You may already have a research journal, either online or on paper, and you may wish to add a new section for blog post ideas. You can also make use of other online note-taking tools, such as Evernote and Google Keep. One of the hardest things about blogging is deciding what to write about – if you are able to write down an idea, and follow this up with some bullet points while you are having lunch, watching Netflix or during your commute, you may find that this greatly helps you when it comes to actually writing your substantive blog post.

3.5 SHARING YOUR BLOG POSTS ON SOCIAL MEDIA

You've done your research. You've carefully crafted your first five blog posts that will be unleashed into the world. You press 'Publish'. But no one is reading. So just how do you get people to actually read what you've written? Well, you need to build up a social media presence and do some other things that will help drive traffic to your blog. Here we'll take a quick look at how to do just that.

In our experience, direct hits on your blog (when a user manually types in a website's address or uses a browser-based shortcut) make up a relatively small number of a blog's total hits. For the LSE's *British Politics and Policy* blog, direct hits were only 13.5 per cent of the total in 2015. The vast majority of the nearly 800,000 total visits were either from searches (41 per cent) or from social media (39 per cent). About 5 per cent were from referrals from other websites and blogs, and 1 per cent were from emails. Figure 3.17 shows how creating a social media presence – external to the blog – can channel readership to it.

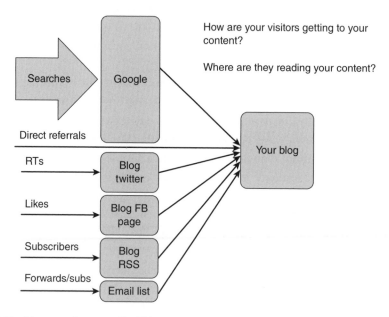

Figure 3.17 How readers can find blog content

So while social media and search engines can be incredibly helpful in driving traffic to your blog, it's worth keeping in mind that they are still acting as filters. Google uses complex algorithms based on the popularity of websites and blogs to steer users towards the content that should best answer their search query (Google, 2016).

3.5.1 Sharing blog posts through social media channels

Most social media platforms use some form of information, be they text or image, to drive traffic to your blog. As such, it's important to think about how your audience will best engage with your content. We find that Twitter, with its more than 300 million users, is an ideal venue for promoting blog content; tweeting the title of your article (especially if it is narrative) or a quote from the piece will help to drive readers to your blog. Some examples of effective tweets could include:

- US elections are seldom about foreign policy. But the FP positions of the candidates matter to the rest of the world + BLOG URL
- 'Academics think that their ideas are complex when in fact they are oftentimes just complexly expressed' + BLOG URL
- Steady growth generates higher levels of wellbeing among citizens than boom and bust cycles + BLOG URL

Facebook gives you more options – you can share more text and images as well. But, you have to decide whether you want readers simply to read a shortened version of your blog post (which may itself be a shortened version of an earlier journal article) on Facebook, or you would rather they were directed to your blog. Remember, if readers click through to the blog post itself, they may remain to view other blog posts on similar topics, something that won't happen via a Facebook posting. Consider as a starting point the groups and networks in which you are already active and look to grow from there. LinkedIn and Facebook can be particularly worthwhile spaces for close-knit, industry-specific interactions. Here are some examples of Facebook or LinkedIn updates that might be instructive:

- This year's election is not likely to mean the end of political gridlock in Washington writes Walter Dean Burnham of The University of Texas at Austin + BLOG URL
- Bridgette Peteet of the University of Cincinnati writes on how graduate school bridge programs can help increase diversity in STEM subject admission + BLOG URL
- What role should social scientists play in society? Louisa Hotson explores the evolution of the social sciences through four periods in the history of political science, each with different implications for how social science makes a difference + BLOG URL

Instagram and Pinterest are primarily based on sharing images – use them when your blog post has an attractive image related to it, such as a chart, graph or photos from your research. As of the summer of 2016, Instagram's captions did not allow for hyperlinked URLs, though users do often use their biography space to add a link to their latest blog article, writing 'see link in bio' in their photos caption. Similarly for Pinterest, the service allows users to click through from their pinned image and go straight to the source from which it was pinned. So, if you are sharing images from your blog post on Pinterest, the user and re-pinner will be able to click through and read. We will of course go into more detail about working with data visualisations in Chapter 4, and photos and videos in Chapter 6.

Hashtags across all of these platforms are also an effective way to reach beyond your individual network. Be sure the hashtag is active enough to reach people but not so populated that your content will get lost in the noise. If the subject of your blog post is a trending topic that day, consider using the relevant hashtags to increase the chances of more eyes on your work. For example, if your blog post about your research on political activism in Brazil is published on the day of the opening ceremony of the Rio Olympic Games, use official and general hashtags about Rio and the Games to take advantage of increased public interest: #Rio2016 #Brazil DilmaRousseff #Rio and #RoadtoRio, for example. More general hashtags about the subject of your blog post or your area of research can also be used: #activism #LatinAmerica #Brazil #favelas and #politics, for example.

Perhaps the most important element of sharing on social media is learning what works. Measuring your efforts and keeping a record of the analytics, even if it is as basic as how many retweets or likes posts are receiving, will help you develop and refine your social strategy. Not only will this recording help you understand the reach and potential impact of your current work, it is also incredibly useful knowledge for your future projects and can save you time and stress down the line when it comes to facing these questions again.

3.5.2 Sharing blog posts by targeted emails

In Chapter 1 we discussed tools that could be used both as social media and for collaboration, such as email and WhatsApp. While addressing a smaller audience than tools such as Facebook and Twitter, emails and messaging services can be a good way to send your blog posts to people in a targeted way. Figures from Campaign Monitor suggest that you are six times more likely to get a click through from an email campaign than you are from a tweet (Beashel, 2014) and for businesses email marketing has an ROI of 3800% (Direct Marketing Association, 2015).

When it comes to building an email list in a professional email marketing tool such as Mailchimp or Campaign Monitor, add and segment your colleagues past and present, research contacts you've met through networking and journalist contacts if you're seeking media coverage. Segmenting the lists means you can send separate emails and invitations to these groups and tweak the messaging and tone accordingly. Be sure not to spam those you haven't met, though, and consider how you frame your invitations to read: do you want feedback? Do you want them to share? Do you want them to comment? Did you cite their research in your piece? Do you think your blog post could feed into their current research? For example, you may wish to send a blog post to your department colleagues, asking them to share amongst their own connections and for feedback, but for the policy team at the office of a legislator that you know has an interest in that area, you could use a different approach. Don't forget to share on relevant mailing listservs. Many blogs and news websites now have a one-click button for easy WhatsApp sharing. Consider connecting to a public WhatsApp group around a relevant topic and sharing a link via mobile.

3.5.3 Sharing blog posts in the 'real world'

Don't forget that the 'real world' exists. Promoting your blog while networking in person can be of great help to your nascent blog. If you are particularly proud of specific blog posts, it can also be useful to print these out in hard copy to give out at events, if you are based in an academic department, or to potential readers who you meet while networking. Be prepared to discuss your work and your experience of blogging at every opportunity: conference presentations, public lectures and festivals are obvious venues to tell an engaging story, but there are plenty of other spaces. Many universities organise informal Show-and-Tells, TEDx talks and Fringe-style events aimed at public audiences. Consider where your findings might provoke discussion and debate and look to see if there are any opportunities for engaging these audiences in dialogue.

Whatever method you pursue, make sure to coordinate a plan for sharing your blog posts. Focus on what specific audiences would be most interested in and would benefit the most from your work and look to build attention from there. In pretty much every case, an audience does not happen overnight. Continue to pursue opportunities where conversations, both online and offline, can occur and make sure to make notes on what works for the future. And, if you are looking for further guidance, Chapter 7 includes more detail on choosing the right platform and planning a coherent social media strategy.

3.6 CONSIDERATIONS

Finishing up our focus on blogs and blogging, there are a few final points to consider.

- There are a few questions you should ask yourself before you start blogging: Who are you trying to reach and do you have the resources to commit the time and energy to creating your own project or individual blog? Since it can take some time to develop a profile, are there existing blogs out there with a built-up audience that you can cross-post or guest blog on in order to connect with the readers you are looking for?
- How will you license your blog work? As the author, you hold copyright and are able to decide if and how it should be shared. If you release it under an open licence, such as a Creative Commons licence, it can be reused and/or remixed.
- Online trolling and other forms of harassment are sadly a part of online life which can be difficult to avoid (Barlow, 2008; Lovink, 2011). While you are unlikely to receive a deluge of comments on your nascent blog – we find that comments tend to increase with readership to some degree – it will still pay in the long run to develop a comments policy for your blog and to display it relatively prominently. Your comments policy is up to you – it can be as lenient or as harsh as you wish. For examples, it's worth looking at the comments policies of the blogs that you follow and adapting their policies for your own purposes. More discussion on navigating online trolling and harassment can be found in Chapter 7.

- How to cite blogs: As blogging becomes a more important form of academic communication, there will be a greater impetus to cite blogs in academic research – something that we encourage. If you wish to cite blogs, we'd encourage the following format, taken from the *LSE Impact* blog:
- Terras, M. (2012) 'The verdict: Is blogging or tweeting about research papers worth it?' *LSE Impact blog*, 19 April. Available at: http://blogs.lse.ac.uk/impactofsocialsciences/2012/04/19/blog-tweeting-papers-worth-it (Accessed 22 May 2012).

3.7 FURTHER READING

Carrigan, M. (2015) 'Why should academics blog about their research? *An answer in pictures.' The Sociological Imagination*. Available at: http://sociologicalimagination.org/archives/17740 [Accessed 14 November 2016].
Accessible introduction to why academics should blog and how it can help them.

Dunleavy, P. (2016) 'How to write a blogpost from your journal article in eleven easy steps.' *Impact of Social Sciences Blog*. Available at: http://blogs.lse.ac.uk/impactofsocial sciences/2016/01/25/how-to-write-a-blogpost-from-your-journal-article [Accessed 14 November 2016].
A good introduction on how to turn your academic article into a short, accessible blog post.

McKenzie, D. and B. Ozler (2014) 'The impact of economics blogs.' *Economic Development and Cultural Change*, 62(3): 567–597. Available via the World Bank at: http://docu ments.worldbank.org/curated/en/388731468336029012/pdf/WPS5783.pdf [Accessed 14 November 2016].
A study that quantifies the impacts of economic blogs on the number of abstract views and article downloads.

Waddington, A. (2015) 'It started with a blog!' *Manchester Policy Blogs*. Available at: http://blog.policy.manchester.ac.uk/posts/2015/09/it-started-with-a-blog [Accessed 14 November 2016].
Summary (including a short film) of how a blog post led two researchers to have input into government policy.

4

CREATING AND SHARING INFOGRAPHICS AND DATA VISUALISATIONS

At their best, graphics are instruments for reasoning about quantitative information. Often the most effective way to describe, explore, and summarize a set of numbers – even a very large set – is to look at pictures of those numbers.

– Tufte (2001: 9)

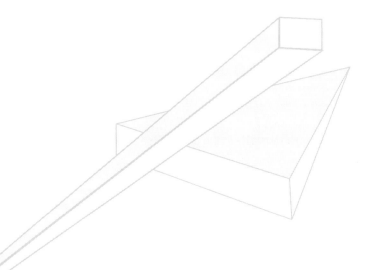

When arriving at Selimiye Barracks in November 1854 to assist with the care of wounded British soldiers fighting in the Crimean War, Florence Nightingale was met with a shocking scene: hygiene and nutrition were so poor that ten times as many soldiers were dying from preventable illnesses such as typhus and dysentery than from battle wounds. 'Our soldiers enlist,' as Nightingale described it, 'to death in the barracks' (quoted in Kopf, 1916: 392). Nightingale immediately introduced new hygiene regimes and kept meticulous records of the causes of soldier deaths. The improved conditions soon dramatically reduced the number of deaths by disease, but back in London the British government was resistant to Nightingale's reforms. This 'had as much to do with the fact that she was a woman in a supervisory role as it did the nature of her position, which was seen as usurping military and medical hierarchy' (Brasseur, 2005: 163).

After the war, Nightingale wrote a report based on the data she had collected, outlining her recommendations for preventing more unnecessary deaths on the battlefield. Despite support from the public, war veterans and even Queen Victoria, her recommendations were ignored and no action was taken (Goldie, 1997: 286). Frustrated, but not deterred, Nightingale created and widely distributed visualisations of the data from the original report, making exceptionally clear to any reader the need for health reform in the Army (Figure 4.1).

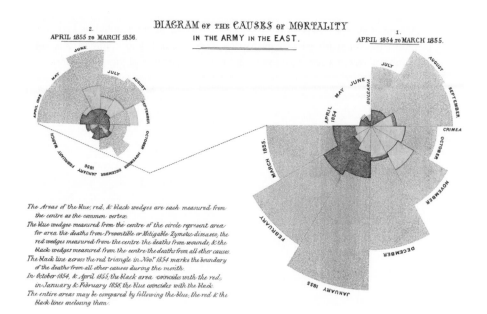

Figure 4.1   'Diagram of the causes of mortality in the army in the East'

Note: Published in Notes on Matters Affecting the Health, Efficiency, and Hospital Administration of the British Army by Florence Nightingale (1858). Printed by Harrison and Sons

Credit: https://commons.wikimedia.org/wiki/File:Nightingale-mortality.jpg This work is in the public domain in its country of origin and other countries and areas where the copyright term is the author's life plus 70 years or less.

Explained with annotations, Nightingale's data visualisations took the form of 'coxcombs' or 'roses', which used segmented discs to show data on the causes of deaths in the Crimean hospital camps each month: preventable diseases in blue, wounds in red and other causes in black. The first rose diagrams showed a picture of mortality in the war from April 1854 to March 1855, contrasted with a smaller circle in the middle representing the mortality rate in one of the most unhealthy towns in England, and pointed readers to the fact that mortality was far too high in the hospital camp context. The second rose diagrams showed that infectious but preventable diseases – not battlefield wounds – were by far the leading cause of death in the camps, with Nightingale dividing the areas within each monthly wedge to show the ratios of cause of death. The third rose diagrams offered a solution to the problems: showing how sanitary improvements at other hospitals had brought about a positive impact on mortality rates.

Nightingale's strategy to present the data as a visualisation highlights her efforts to 'make complex data clear to a resistant audience' (Brasseur, 2005: 175). The data visualisations had the intended impact: Nightingale's recommended reforms were understood, accepted and carried out without exception. Regarded as 'a true pioneer in the graphical representation of statistics' (Cohen, 1984: 132), Nightingale was elected the first female member of the Royal Statistical Society in 1859 and later became an honorary member of the American Statistical Association.

Whether you are an astrophysicist looking to share research on stellar structure with students, a nurse looking to start discussions around ward hygiene with colleagues, or a criminologist looking to get journalists and policymakers interested in your research on reoffending patterns, using infographics and data visualisations as part of your communication strategy can be an exciting and effective way to reach your audience. This chapter looks at some definitions of infographics and data visualisations, and provides a brief history of their use, before examining the ways in which they can create impacts in a variety of settings. We look at how you can create your own data visualisations and infographics using some of our favourite tools and with tips from the professionals.

4.1 DEFINING INFOGRAPHICS AND DATA VISUALISATIONS

A data visualisation refers to any graphic that displays information in an understandable visual format. Data visualisations like graphs and charts often deal with very large amounts of data and enable us to draw conclusions much more easily than if all the data were presented as words or in a table. Data visualisations can present a dataset with minimal decoration, though digital data visualisations tend to be interactive and allow users to explore and share on social media different areas of the dataset.

An infographic is a type of data visualisation, but has a distinct story or editorialised narrative. Infographics are often designed to tell a particular story which has been drawn out of a dataset, aimed at a specific audience. An infographic might include images, icons, illustrations, charts and attractive design features, but doesn't usually include a full dataset. Long, vertical illustrated narrative infographics are commonly used by advertisers to promote their products through content marketing, but researchers and academics can also harness this accessible and attractive style in order to maximise quick reader understanding of complex material.

Another distinction is between the two approaches to the graphic representation of information. A more minimal, 'no frills' approach that defines most academic-produced data visualisations is represented in the influential work of Yale University statistics professor Edward R. Tufte, whose works, including *The Visual Display of Quantitative Information* (2001), are recognised as a *tour de force* in information design. Tufte's philosophy goes that any graphic elements of a design that do not communicate specific information, or are purely decorative, are not necessary and should therefore be omitted. For Tufte, these unnecessary design elements and distractions are 'chartjunk' (2001: 107) and only decrease the value of what has been produced. This approach to data visualisation is taken up mostly by scientists and academics in academic journals, writing for a serious audience with time to investigate all possible conclusions in front of them.

But can't data visualisation be beautiful as well as functional? At the opposite end of the spectrum to Tufte are those who support the strong use of illustration and decoration in information design. Graphic design work created for *Time*, *New Scientist* and *The Guardian* is perhaps some of the most influential in this approach. Here, decorative elements and visual metaphors reinforce and enhance both the narrative and reader understanding, rather than detract from it. Information designer Georgia Lupi summarised this approach perfectly in the FiveThirtyEight data visualisation podcast *What's The Point*: 'Beauty is a very important entry point for readers to get interested about the visualisation and be willing to explore more. Beauty cannot replace functionality but beauty and functionality together achieve more. Beauty is an asset' (Avirgan, 2016). Also taking this approach is David McCandless, who notes in the highly popular *Information is Beautiful* that our information-rich world has made us visual consumers and learners: 'what we need are well-designed, colourful and – hopefully – useful charts to help us navigate' (McCandless, 2009: 6).

TWO APPROACHES TO DATA VISUALISATION DESIGN

Georgia Lupi: 'Beauty is a very important entry point for readers to get interested about the visualisation and be willing to explore more. Beauty cannot replace functionality but beauty and functionality together achieve more. Beauty is an asset.' (Lupi in Avirgan, 2016)

Edward R. Tufte: 'The purpose of decoration varies – to make the graphic appear more scientific and precise, to enliven the display, to give the designer an opportunity to exercise artistic skills. Regardless of its cause, it is all non-data-ink or redundant data-ink, and it is often chartjunk.' (Tufte, 2001: 107)

But should infographics and data visualisations completely replace full data tables or long and winding research write-ups? Mikko Jarvenpaa, CEO of online data visualisation tool Infogram, shared these words with us:

Visualized data is engaging data. Because it works as a visual metaphor, data visualization triggers a different cognitive process than reading about the same data in text format. First, having more than one way for the audience to approach the information is beneficial for capturing their interest. Second, having the two modes of presenting the data – the visual and the textual – is beneficial for ensuring the audiences understand and retain the information. Depending on the individual and the situation, some information is better ingested in a visual format and some better in a textual format. Researchers should not be concerned about redundancy or repetition in using both in their research: this is actually doing the audience a favour. (Personal correspondence, April 13 2016)

What both infographics and data visualisations have in common – regardless of their design approach – is the visual representation of information or data in an attractive or more accessible format. They are often used 'to convey complex information to an audience in a manner that can be quickly consumed and easily understood' (Smiciklas, 2012: 3), and can be 'informative, expressive, persuasive, and more or less rhetorical, depending on [their] design, context and audience' (Bestley, 2013).

4.2 INFOGRAPHICS AND DATA VISUALISATIONS: A HISTORY IN RESEARCH COMMUNICATION

The buzz around infographics and data visualisations on social media may lead us to think of them as something of a modern phenomenon. But the popular infographics we read, share and create today have roots in much older scientific visualisation techniques and research communication efforts going back hundreds of years.

4.2.1 William Playfair invents line, bar and pie charts

Scottish engineer and political economist William Playfair (1759–1823) is one of the founding talents of modern graphical design: inventor of the time-series line graph, area chart, bar chart, pie chart and circle graph. The publication of his *Commercial and Political*

Atlas in 1786 contained 44 charts and graphs summarising trade between England and other countries, with accompanying annotations, tables and observations about national finances. Playfair's visual representation of economic data in the *Atlas* was a first for this era, with Playfair reasoning:

> Information, that is imperfectly acquired, is generally as imperfectly retained; and a man who has carefully investigated a printed table, finds, when done, that he has only a very faint and partial idea of what he has read; and that like a figure printed on sand, is soon totally erased and defaced. (1786: 3)

Playfair went on to publish *Statistical Breviary* in 1801, containing what is widely acknowledged as the first pie chart, as part of information on populations and military power across Europe. The close-up in Figure 4.2 shows the land areas of the Asiatic, European and African portions of the Turkish Empire. Playfair coloured the Asiatic segment green to signify a maritime power; the European red to show a land power; and the African yellow. The result is 'the first pie chart to display empirical proportions and to differentiate the component parts by colour' (Spence, 2005: 355).

Playfair's charts were 'in his own day neither praised nor imitated' (Reid, 2006: 274) and it would be another hundred years before scientists would embrace pictorial representation. Modern statistical graphs are almost identical to those published by Playfair, and the visualisations in the *Atlas* contain all of the elements that we take as essential in time-series graphs now: graduated and labelled axes, grid lines, a title, labels, lines indicating changes in the data over time, colour to categorize the different time series and accumulated quantities (Spence, 2006: 2427).

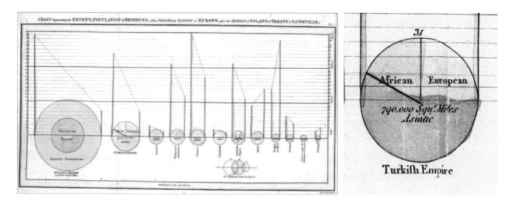

Figure 4.2 Pie charts and close-up from William Playfair's *The Commercial and Political Atlas* (1786) and *Statistical Breviary* (1801), with close-up showing the proportions of the Turkish Empire located in Asia, Europe and Africa before 1789

Credit: https://en.m.wikipedia.org/wiki/File:Playfair_piecharts.jpg and https://en.m.wikipedia.org/wiki/File:Playfair_piecharts.jpg. This work is in the public domain in its country of origin and other countries and areas where the copyright term is the author's life plus 100 years or less.

4.2.2 Charles Booth's poverty maps change the way we understand demographics in London

Charles Booth's maps of poverty levels in Victorian London have been highly influential in shaping how statistics can be presented through maps (see Figure 4.3). Published in 17 volumes between 1889 and 1903 as *Life and Labour of the People of London* (1889, 1902a, 1902b), social reformer Booth's intricately detailed maps showed the results of survey data about levels of income from over 120,000 households across inner London (Dorling et al., 2000). With streets and buildings coloured to correspond to social class, similar classes were given similar colours so that general trends across the capital could be made more apparent. Yellow indicated the most wealthy upper classes; red and pink the comfortable middle classes; light blue the poor and casual labourers; and black, in Booth's words, 'the lowest class; vicious; semi-criminal'.

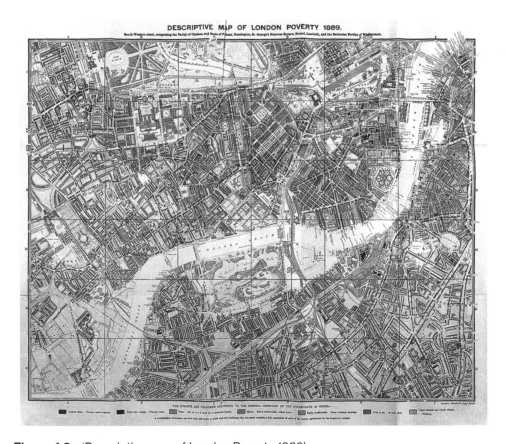

Figure 4.3 'Descriptive map of London Poverty 1889'

Note: 'South Eastern sheet, comprising the registration districts of St. Saviour's Ward and St. Olave's, Southwark, and parts of Lambeth, Camberwell, and Greenwich' from Charles Booth's *Life and Labour of the People in London* (1889)

Booth's research – carried out with a team of volunteers including economist Clara Collet and LSE co-founder Beatrice Webb – was triggered by a desire to counter the sensationalised reporting of rioting in London during February 1886, the causes of which were wrongly attributed to a dangerous social revolution rather than poverty, with politicians looking to policies of 'suppression rather than understanding' (Bales, 1999: 153). Booth recognised the value of evidence in these discussions, and his maps were an effective addition to public debates about poverty, crime and society during this era. Whereas Nightingale's rose diagrams had 'heighten[ed] the tone, intensity and drama through the diagrams' visual rhetoric' (Potter, 2010: 87), Booth's maps paradoxically downplayed the era's emotive visual rhetoric of poverty in order to make it 'seem a problem that could be addressed, rather than an insurmountable crisis' (Kimball, 2006: 353).

Booth 'became what might have been the first sociological "household name"' (Bales, 1999: 153), with his findings widely reported around the world. Looking at the influence of his designs on cartography and social science research, his poverty maps are 'a seminal work in the history of London maps and the development of geodemographics' (O'Brien and Cheshire, 2011).

4.2.3 Vienna's Isotype brings 'dead statistics' to life

A different take on the visual representation of data is the use of pictograms: the pictorial representation of statistics. In 1920s Vienna, sociologist Otto Neurath founded the social and economic museum of Vienna. He chose to use strikingly simple pictograms inside, enabling a wider understanding of social, technological, biological and historical ideas. Working together with illustrator Gerd Arntz and mathematician Marie Reidemeister, Neurath developed the International System of TYpographic Picture Education – or Isotype. In the context of a city still young in its democracy and with low levels of literacy, Neurath sought to 'give fundamental, strictly scientific information for social understanding, even to the less educated, without depressing them in the way learned books and statistical tables do' (Neurath, 1973: 217). For Neurath, the aim was to 'represent social facts pictorially' and to bring 'dead statistics' to life by making them both attractive and accessible (Burke 2009: 210).

Arntz, Reidemeister and Neurath were influenced by the work of William Playfair, but the Isotype's 'unique grammar and syntax were developed only gradually' (Jansen, 2009: 227). The symbols themselves are packed with meaning and emotion, characterising social concepts with subtle accuracy. A famous example is the symbol for the unemployed: a sombre, hunched man with both hands in his pockets. Each Isotype symbol always represented the same concept – lending stability and continuity to the graphics – and a greater quantity would always be represented by a greater number of signs, not by varying the size of a single sign.

Looking at their own influence on the world, the Isotype has long inspired designers, artists and politicians, with Jansen (2009) identifying its influence in Otl Aicher's 1972 Munich Olympic Games pictograms (Figure 4.4) and in data visualisation in the *New York Times*. Paul Mijksenaar (1997: 30) has argued that Isotype has 'contributed to the development of pictograms and, indirectly, of the "infographics" that are becoming increasingly popular'.

The rise of Austrian fascism forced the trio to flee to the Netherlands where they set up the International Foundation for Visual Education in The Hague. Marie and Otto fled again from German invasion, this time to England, where they established the Isotype Institute in 1942. From here, Isotype was applied to wartime publications sponsored by the Ministry of Information. Neurath's oft-quoted phrase has never been more poignant: 'Words divide, pictures unite' (1931/1991: 569).

Figure 4.4 Otl Aicher's pictograms on display for the Olympic Games in Munich, Germany in 1972, symbolising curling, figure skating and ice hockey

4.2.4   Internet + open data = infographics overload

Bringing us up to the present era is Sir Tim Berners-Lee and the invention of the World Wide Web in 1991, which facilitated the wide creation and social sharing of data visualisations and infographics outside of academia and journalism. For many decades, infographics and data visualisations had traditionally been the domain of 'an educated, knowledgeable, and skilled group of individuals' (Lankow et al., 2012: 30–31), produced almost exclusively by academics. Journalists too had made wide use of the format in printed newspapers and magazines – with Pasternack and Utt noting as early as 1990 that infographics had become 'the rule rather than the exception in a large number of daily newspapers'

(1990: 28). Yau (2010) marks 2007 as something of a turning point for public interest in creating, consuming and sharing information in this format, with Google showing large jumps in searches for infographics from this point, and again in 2012 (see Figure 4.6). Around 2008 we also see the start of a new design trend for infographics: the emergence of the elongated, vertically oriented format suitable to fit blog platforms such as WordPress and Blogger. Infographics in this format, such as the Working Lands for Wildlife graphic in Figure 4.5, were popular for commercial marketing purposes, as organisations sought to find new ways of sharing their products or key sales information. 'The long, skinny graphic, designed to fit within a blog's width, became ubiquitous and almost instantly synonymous with the term *infographic*', note Lankow et al. (2012: 31, original emphasis).

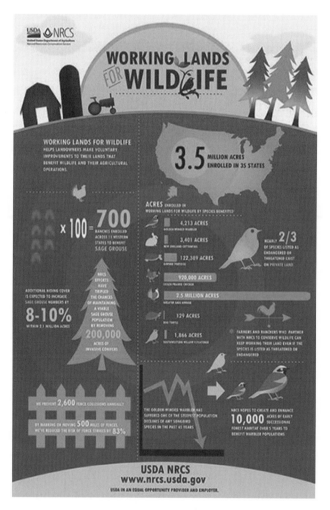

Figure 4.5 Working Lands for Wildlife infographic from USDA NRCD, South Dakota

Credit: www.flickr.com/photos/nrcs_south_dakota/9788793604 (CC-BY-SA 2.0)

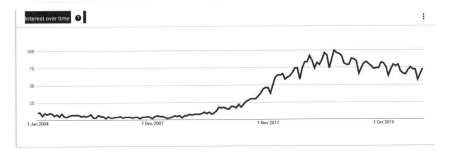

Figure 4.6 Google Trends results for 'infographics', showing interest over time in web searches from 2004 to 2016

Credit: www.google.com/trends/explore?date=all&q=infographics

This increased public interest can be connected to the rise of personal, lifestyle and academic blogs, as well as citizen journalism and science communication initiatives. The push to make large government and other datasets open online has only aided the phenomenon (Fernandes et al., 2013: 86; Or, 2013: 4839). An example of this is the US federal government's data.gov initiative, which 'includes over 3,120 datasets ranging from national education and public health to environmental and transportation data' (Diakopoulos et al., 2011: 1717). New visualisation hubs such as *The Guardian*'s datablog (theguardian.com/data) have also sprung up to fulfil public interest, while the award-winning Graphics Department at *The New York Times* is recognised as 'setting the benchmark for editorial infographic content … widely considered the best in the world at crafting data-driven content' (Lankow et al., 2012: 122).

From 2012, online tools and apps that allowed for the easy creation of infographics and data visualisations by anyone had become popular. Contrasting with the unwieldy nature of Photoshop, the likes of Piktochart, Visme, Visual.ly and Infogram were beginner-friendly and provided ready-made templates, allowing novice users in their millions to input data from a spreadsheet, add text or logos, and undertake some of the whizzier editing previously only done by professionals. One-click sharing buttons meant that infographics and data visualisations could be shared on social media in seconds. Social media platforms now widely privilege the visual, encouraging more engagement from users and their connections.

This opening up of the creation of infographics and data visualisations to the wider public through online tools is not without its critics, as some have argued that the overall quality of material has gone down and often not enough attention is paid to referencing sources, using responsible messaging and using design features wisely. But this is true for any medium recently opened up to the wider public – and in our context of research communication, we feel that readers and creators will be more aware of these factors than most.

The only place that we can end our overview of the history of data visualisation is with two major political events from 2016 that saw pollsters and their data visualisations get it badly wrong: the UK voting to Brexit from the European Union on 23 June and the USA voting to elect Donald Trump as their 45th President on 8 November. The run-up to both events saw wide media, public and academic use of data visualisations and polling. Public interest and social media sharing of these election prediction visualisations was high, signalling to some the ever strengthening power of data visualisation and particularly election forecasting. But pollsters came in for heavy criticism after incorrectly predicting a comfortable win for the Remain campaign in the UK and a landslide win for Hillary Clinton in the US. In both cases, those on the political left were shocked as viral data visualisations from respected sources like Nate Silver's *FiveThirtyEight* were proven wrong again and again as the final results were revealed.

There were of course many factors that influenced the failure of the predictions, including a reliance on telephone surveys over online surveys (Clarke et al., 2016), and a denial by the left and the media elite of the concerns of those voting out of the EU. It is too soon to tell what effect these failures will have had on public interest in the consumption and sharing of data visualisations, but if the rejection of evidence-based opinions in the run-up to these political events is anything to go by, it's going to be more important than ever for those working in research communication to do it well, do it fully and do it alongside informed analysis.

4.3 WHY INFOGRAPHICS AND DATA VISUALISATIONS ARE USEFUL ACROSS THE RESEARCH LIFECYCLE

Infographics and data visualisations have undertaken a journey through the battlefields of the Crimean War, via the design influences of the twentieth century, past the bleeps and whirrs of dial-up internet, and now to the social media feeds we scroll through on our tablets and smartphones. Their popularity is clear and it is easier to create them than ever before. But how do we know if they have any impact for research and academic settings today?

Jason Lankow, Josh Ritchie and Ross Crooks, in *Infographics: The Power of Visual Storytelling*, argue that 'the visualization of information is enabling us to gain insight and understanding quickly and efficiently, utilizing the incredible processing power of the human visual system' (2012: 12), and Helen Kennedy (2015), in her article for *LSE Impact Blog* on factors which affect our engagements with data visualisations, observes that 'for a long time, experts have argued that visualisations are important tools for making data transparent and for communicating in ways that data themselves cannot'.

WHAT SHOULD WE CONSIDER WHEN USING INFOGRAPHICS?

Writing for the *LSE Impact Blog* about her project Seeing Data (http://seeingdata.org/), Helen Kennedy (2015) identified a range of factors which affect our engagements with data visualisations:

Subject matter: when the subject matter speaks to our interests, we're more likely to engage with data visualisations.

Source/location: when visualisations are encountered in already-trusted media that we view or read regularly, we are more likely to trust them; otherwise, we don't.

Beliefs and opinions: we like visualisations which communicate data in a way that fits with our world views, but some of us also like our beliefs to be challenged by data and visualisations.

Time: engaging with visualisations can be seen as hard work by people for whom doing so does not come easily, so having time available is important in determining whether we want to do this 'work'.

Visual elements: the conventions that visualisation designers draw upon play a role in determining whether we're willing to spend time looking at a visualisation. These may appeal to us, or they may seem too unfamiliar and therefore offputting.

Emotions: visualisations provoke emotional reactions – if we feel immediate confusion about a visualisation, we are less likely to invest time and effort in making sense of it. Subject matter, visual style and other factors all provoke emotional reactions.

Confidence: we need to feel confident in their ability to make sense of a visualisation, in order to be willing to give it a go. This usually means feeling like we have some of these skills:

- **Language skills**, to be able to read the text within visualisations (not always easy for people for whom English is not their first language).
- A combination of **mathematical or statistical skills** (knowing how to read particular chart types or what the scales mean) and **visual literacy skills** (understanding meanings attached to the visual elements of datavis) – sometimes called '**graphicacy' skills**.
- **Computer skills**, to know how to interact with a visualisation on screen, where to input text, and so on.
- **Critical thinking skills**, to be able to ask ourselves what has been left out of a visualisation, or what point of view is being prioritised.

Looking at the literature on infographics and data visualisations created professionally for research dissemination, there are many examples of these outputs being used to successfully increase user engagement with and understanding of research findings – both in traditional printed formats and via social media, right across the Research Lifecycle.

4.3.1 Infographics and data visualisations can help make policy impacts

When the British Cycling Federation released its 2014 report on how government investment in cycling could bring in billions of pounds' worth of savings and benefit the health of cyclists and non-cyclists alike, it chose to use an infographic to grab the attention of the intended audience of policymakers and politicians. Rather than overwhelm readers with complex forecasts, the report opened with an eye-catching full-page infographic (see Figure 4.7). The infographic is simple, attractive and effective, walking the reader through eight bitesize take-aways on what further investment in cycling could bring to the UK, from saving the NHS £17 billion to increasing the mobility of the nation's poorest families by 25 per cent.

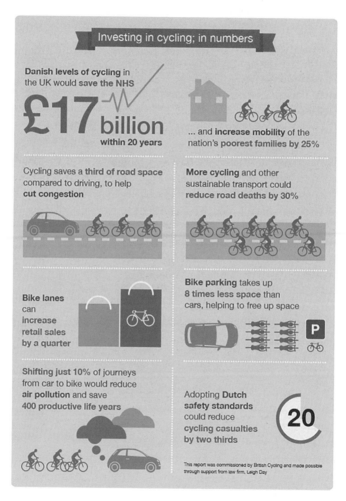

Figure 4.7 'Investing in cycling; in numbers' infographic, from British Cycling. Included in Dr Rachel Aldred's report 'Benefits of investing in cycling' (2014)

Credit: Reproduced with permission from British Cycling

2 March 2015 announced an investment of £114m into cycling infrastructure in eight cities, referencing three of the eight infographic take-aways from Aldred's British Cycling report (2014). The infographic was also reproduced by Highways England for use in its cycling strategy, and has also made it into the government's 2016 Cycling and Walking Investment Strategy.

Aldred told us:

> In terms of the impact of the report, one thing I hoped to achieve was to encourage the debate to focus on the benefits of investing in cycling – i.e. encouraging policy-makers rather than individuals to take action. It was sent to all MPs and I think that British Cycling promoting it, and making it look so striking, was very helpful in getting the message across. I've certainly heard some of the figures quoted back at me when I've been speaking at events at TfL [Transport for London] and elsewhere. I think the infographics have been brilliant, I keep seeing them appear on social media. I'm an infographic convert – I think they are a great way to get ideas across.

This example shows us that with targeted and well-timed promotion, accessible and attractive design, and being paired up with the launch of a full report, infographics can help research make policy impacts.

4.3.2 Infographics and data visualisations can provide evidence-based arguments to challenge dominant media rhetoric

> We were counting bodies on the shores, writing a press release, getting no media coverage, then doing it again days later. … We knew we would have to change the narrative behind events before being able to change policy.

Speaking at the Polis Conference on journalism and crisis at the London School of Economics in April 2016, Andrea Menapace describes how the Open Migration project (http://open migration.org) decided to build data visualisations into the heart of its dissemination and engagement strategy. Open Migration, created by the Italian Coalition for Civil Liberties, aims to extend the protection of migrants' human rights and raise public awareness through sharing policy analysis and personal stories. After having little success using op-eds to promote their work, they realised that in a media landscape already saturated with negative narratives about migration – with refugees painted as dangerous terrorists at worst and greedy economic migrants at best – there would be next to little interest in their outputs. 'It was a toxic media narrative', says Menapace, 'and no matter how passionately we wrote and how strongly we argued, nobody was paying attention. We weren't going to be able to make a policy impact with our work when the media was not even giving us a chance.'

The Open Migration team got thinking about ways to continue challenging the dominant negative media narratives around the migration crisis, and started experimenting with adding infographics and data visualisations to their work. They also strived to put into context the emergency and body counting narratives being discussed, by sharing personal stories and policy analysis. The example in Figure 4.8 is one of many produced as part of a fact-checking series, in which the speeches of politicians or the columns of journalists are fact-checked against openly available data. Here, the words that Claudio Cartaldo (2015) wrote in *Il Giornale*, referring to an invasion of Muslims into Italy, was found by Open Migration to be 'mostly false': when comparing data on the religion of foreign citizens of Italy from 1993 to 2014, the percentage of Muslims had remained unchanged at 32 per cent.

The data visualisations contained the same information they were writing about as before – data collected from sources such as Eurostat and the UNHCR. But after sharing

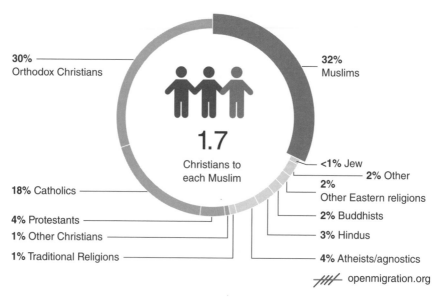

Figure 4.8   Religion among foreign citizens in Italy, 2014

Credit: Open Migration, from Alessandro Lanni's blog post 'How to debunk the myth of a Muslim invasion in Italy'. https://openmigration.org/en/fact-checking/how-to-debunk-the-myth-of-a-muslim-invasion-in-italy

these regularly through their social media channels and other NGO networks they saw a sharp increase in traffic and social media engagements. Building on this appetite for more visual and socially shareable content, they created an infographics hub on their website which housed all the visualisations in one place (http://openmigration.org/en/infographics/#all). What they saw was a marked increase in their infographics being used online in debates and discussions.

This example shows us that by finding regularly updated and reliable data sources, thinking about the most effective ways to share your key message, and creating an attractive space for your readers to explore, infographics can contribute to the challenging of dominant media narratives.

Project member Giulio Frigieri finished his Polis Conference presentation with a solemn reminder that although we may marvel at the beauty of infographics and enjoy sharing them, 'behind every data point, somewhere out there is a human being'.

4.3.3 Infographics and data visualisations can transcend language barriers to help spread vital information

In a context where the poor formatting of medical information and its often too complex presentation has led to many patients not attaching value to the written information they receive, health researchers and practitioners are increasingly creating infographics and pictograms for patient leaflets on identifying symptoms of illness and managing the side-effects of drugs (see Basara and Juergens, 1994; Raynor et al., 2007). Dowse et al.'s (2010) research into developing visual graphics for communicating complex information about antiretroviral side-effects to the South African Xhosa population found that visual images or pictograms which 'give an accurate lifelike representation of the human body; and use the expressive power of the human body to construct meaning through different body postures, arrangements, and facial expressions' have been shown to improve patients' knowledge, particularly in low-literate groups, when used as a counselling tool or included in information leaflets (2010: 222, 213).

Where low literacy skills can disempower patients in healthcare clinic interactions and text-based patient information materials are written in too complex a style, the researchers found that uninformed patients were more likely to make decisions that negatively impacted the effectiveness of their treatment. Aiming to design visuals or pictograms illustrating various antiretroviral side-effects in order to overcome the barrier of inaccessible written material, the team worked closely with a graphic artist, sharing ideas in workshops with Xhosa groups. The pictograms shown in Figure 4.9 – simple in style and not overly design-oriented – were understood by users to varying degrees, but user group testing showed 100 per cent comprehension of the visual metaphor pictograms for nausea and high rates for three of the most common symptoms. Recommendations from the study include that infographics or similar pictogram visuals should be designed in collaboration with the target population and a graphic artist, should take cognisance of the audience's literacy skills and culture, and should include stages for modification and evaluation. This study demonstrates how useful visual metaphors of information can be for communicating with groups with low levels of literacy or across language barriers.

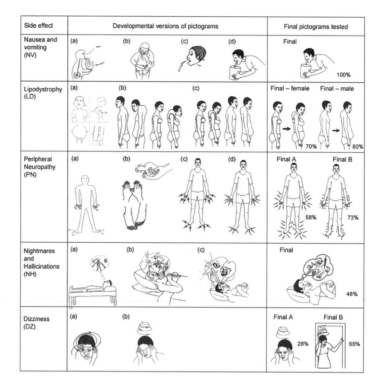

Figure 4.9 Successive versions of the pictograms and the percentage of participants making the correct interpretation of the final version

Note: Only the final images were quantitatively tested.

Credit: From Ros Dowse, Thato Ramela , Kirsty-Lee Barford and Sara Browne (2010) Developing visual images for communicating information about antiretroviral side effects to a low-literate population, *African Journal of AIDS Research*, 9:3, 213–224, DOI: 10.2989/16085906.2010.530172

4.3.4 Infographics and data visualisations can make sense of multiple unwieldy open datasets

In Brazil, where dengue fever affects more than 1.5 million people each year, a research team from prominent think tank Fundação Getulio Vargas (FGV) used infographics to create a dengue panorama of Brazil, bringing public, media and political awareness to the epidemic.

In 2016, one of the major projects of FGV's Department of Public Policy Analysis (FGV/DAPP) centred on public health, involving the organisation, consolidation and publicising of different data sources about dengue fever, an epidemic in Brazil transmitted by the Aedes Aegypti mosquito. The work was coordinated by Professor Marco Ruediger, sociologist and FGV/DAPP's director, and researchers Janaina Fernandes and Wagner Oliveira. Oliveira told us that people in Brazil had been

> especially concerned about this issue recently due to both the Olympic Games in Rio de Janeiro and because of the rise of two new viruses transmitted by the same mosquito, causing new (and almost unknown) epidemics: Zika and Chikungunya Fever. The aim of FGV/DAPP's work is to shed light on the debate using data visualization in order to make it easier for people to: find where cases of dengue are registered, to compare different regions, reveal the need for more attention in resource allocation according to epidemiological data, and therefore discover how the public budget is being spent in terms of combating the vector. (Personal Correspondence, May 9 2016)

As discussed in the previous section, more datasets are available to the public than ever before but problems exist in comprehension, how to access them and in finding solutions to incomplete datasets or datasets that don't match up. FGV/DAPP encountered this problem. Publicly available data provided by the Ministry of Health (Federal Government), the State Department of Health and the Municipal Department of Health for São Paulo and Rio de Janeiro had to be consolidated first. Oliveira told us that 'the team of researchers found that each level of government used a different methodology to count and publish epidemiological data, something that justified even more the need to make these efforts to consolidate different sources of data'.

From there, the team used an online data visualisation tool – Infogram – to create data visualisations that made clear the true impact of the numbers in an engaging and impactful way. FGV/DAPP was able to create interactive charts and infographics to accompany their more detailed research reports, along with separate charts to share with the public and the media (see Figure 4.10). Mapping the data in a visual format allowed them to identify a neighbourhood with a very high number of cases that was previously overlooked, then send in drones to investigate the mosquito population and incidences of dangerous stagnant water sources. For Oliveira, 'the impact was enormous: not only were we able to find many highly possible infectious sources, but also local residents began to respond to the presence of the drones and voluntarily offered their help to find trouble spots in their neighborhood'.

Their strategy of pulling together the datasets into an accessible mini series of infographics had the intended impact: FGV/DAPP attracted the attention of the biggest media group in Brazil and appeared on the news for a whole day's specialist programming, with interviews, infographics updated live and commentary on the results of the drone flight and findings (Figure 4.11). The media coverage even resulted in an official comment from the Brazilian Ministry of Health about its efforts to allocate resources to combat dengue in 2016.

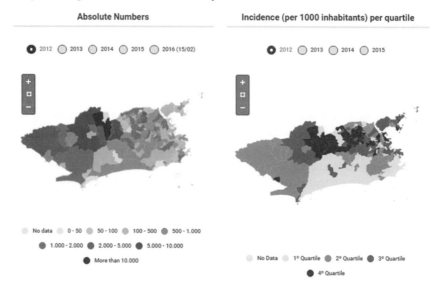

Figure 4.10 Part of FGV/DAPP data visualisation showing the five neighbourhoods with the highest number of cases of dengue in Rio de Janeiro city

Credit: Reproduced with permission from FGV/DAPP and Infogram. Source and full infographic: https://infogr.am/_/3HTaPwlEr0txYWUhgWYh

Their work had even more impact. By comparing the datasets, the team found inconsistencies with the way resources were allocated. Oliveira again:

> We found that the state of Goiás was receiving four times less money per capita than the Federal District, despite having an incidence of dengue that was eight times higher. It was decided to take this information to the relevant authorities, including the State Governor of Goiás. This created a lot of discussion about the parameters used for health budget transfers between different administrative spheres.

This section has given us a flavour for how data visualisations and infographics have been used effectively by academics, scientists and researchers all over the world in order to communicate their research in a public setting. Some have used social media, some have used traditional media and others have taken place in purely research settings. Whatever the setting, you can learn from them as you start to pull together your own ideas about how data visualisations and infographics might work for your own project, as we will discuss in the next section.

Figure 4.11   Image of the FGV/DAPP data visualisations being discussed on GloboNews in Brazil

Credit: Globo News and FGV/DAPP. The full programme can be watched at http://g1.globo.com/globo-news/jornal-globo-news/videos/v/estudo-da-fgv-monitora-areas-com-focos-do-mosquito-no-rio/4842466

4.4   HOW YOU CAN CREATE SUCCESSFUL INFOGRAPHICS AND DATA VISUALISATIONS FOR YOUR PROJECT

Up to this point we've covered the history of data visualisations and infographics, explored the impact of their use in a variety of settings, and now we will move on to how to create them for your own project.

4.4.1   The questions to ask before you create anything

Before designing anything, ask yourself some key questions about your audience and what you want your data visualisations to achieve. Not only will this give you space to think critically, but you will have a plan to return to at future stages and make sure your designs are doing their job as you originally envisioned.

The questions to ask before you create anything:
Question 1: What data do you have access to?
Question 2: What story do you want to tell and what do you want to achieve?
Question 3: Who are your target audiences?

Question 1: What data do you have access to?

If you are a PhD student or researcher in the field collecting data, then you will already have an idea of what sort of data you can use in your visualisations, be it qualitative or quantitative. If data collection is not part of your project or you want to experiment first, then there are many open databases from which you can download datasets. The following websites all provide country-specific data on topics like the economy, demographics, education, science and technology, and social development: Google's Public Data Explorer (google.com/publicdata/directory), Eurostat (ec.europa.eu/eurostat), UNESCO Institute for Statistics (uis.unesco.org), The World Bank (data.worldbank.org) Nation Master (nationmaster.com), US Government Open Data (data.gov), The Open Science Data Cloud (opensciencedatacloud.org), Google Scholar (scholar.google.com) and the World Health Organisation (who.int).

Make sure that all the information you are going to use in your infographic is from a credible source and can be shared in the public domain. If you are collecting the data yourself, then make sure you ask the permission of participants to use their data and answers in this way – particularly for qualitative data or when working with vulnerable or high risk groups.

Question 2: What story do you want to tell and what do you want to achieve?

Now you know what raw or unfiltered data you have to work with, start to explore what stories you can tell with it. Whether you are working with data from your project or you are using data from another source, it is important to work through it and thoroughly understand it: accidental – or intentional – distortion of the data needs to be avoided at all costs. Find those little golden nuggets of trends and stories before you start with any design work. The story should dictate the design, not the other way around. Identify the most compelling parts of your dataset but remember that it is best to find something simple and strong. Or, as Tufte so succinctly puts it, 'a puny data set cannot be rescued by a graphic (or by calculation), no matter how clever or fancy' (2001: 15).

An important consideration when selecting stories to tell in your data visualisation is around the ethics of representation. Feminist standpoint theorists would argue that no knowledge – or dataset or visualisation – is neutrally created; it is socially situated and arguably permeated with a history 'tied to militarism, capitalism, colonialism, and male supremacy' (Haraway, 1991: 188). The perspectives of underrepresented and oppressed groups have traditionally been systematically excluded from what we might think of as 'general' knowledge, and this consideration should come into which stories you tell through the datasets you choose to use, and in which stories you bring through into your own visualisations. Who collected the data? Who funded its collection? Whose labour happened behind the scenes and under what conditions? Whose stories are you not telling? Catherine D'Ignazio, an Assistant Professor of Data Visualization and Civic Media at Emerson College, who investigates how data visualisation, technology and new forms of storytelling can be used for civic engagement, sums up this important point:

> While there is a lot of hype about data visualization, and a lot of new tools for doing it ... fewer people are thinking critically about the politics and ethics of representation. This, combined with a chart-scared general public, means that data visualizations wield a tremendous amount of rhetorical power. Even when we rationally know that data visualizations do not represent 'the whole world', we forget that fact and accept charts as facts because they are generalized, scientific and seem to present an expert, neutral point of view. (D'Ignazio, 2015)

Question 3: Who are your target audiences?

Finally – and perhaps most importantly – you should consider who your audiences might be and what level of comprehension or expectations they have. This will influence how detailed a story will be or what aspect of it you want to tell. Is your audience broad – for example, all internet users – or more specific – for example, all internet users living in Brazil – or very specific – internet users living in urban Brazil who are interested in the topics of healthcare and reproduction? Think about who will be interested in your visualisation and your research, and map these out: the more specific, the better. If you are looking for more guidance on how to identify direct and indirect audiences for your work, Chapter 7 goes into more detail.

Now is also the time to think about the aims of your visualisation: what do you want your audience to do with the information they learn from your visualisation? Perhaps you are looking to put a new spin on an overlooked topic or persuade people to adjust their behaviour? Do you want to raise awareness of your wider research? Do you want media coverage? Consider whether the data story that you've identified has the potential to do this, and whether your audience will be interested enough to engage with it. You can also start to note ideas about the design elements of your visualisation and how this relates to your audience, for example what colours to use, traditional or modern stylings.

4.4.2 Different types of data visualisations and infographics for you to design

Covering map-based visualisations, word-based visualisations and various types of charts and graphs, we will now outline the best use for each and include real-life examples. Each of these elements can be used on its own as a separate data visualisation or as part of a larger narrative infographic, and each can be created with more or less of a focus on additional design elements depending where you (and your audience) fall on the beauty/functionality debate.

Maps

- Maps are, somewhat obviously, the best option for showing trends geographically. Maps are ideal if you're looking for a format that can quickly show that a phenomenon is unique to one area, either on a global or regional scale.
- Map visualisations are often used effectively in psephology, geography and health studies, and can be produced as static images or interactive maps that allow users to click and see information relevant to their region or over different time periods. Both formats are popular with media outlets and in social media sharing.
- Choropleth maps use colour to show readers information about population density or per-capita income in different regions, for example. A colour scale is assigned to categorical or numerical data and the value for each region is used to colour the region on

Figure 4.12 Map of the results of the 2014 EF English Proficiency Index

Credit: Produced by EF Education First (CC-BY-SA 4.0)

the map. Choropleth maps are based on predefined political borders, such as states or counties, and are most effective when the viewer can easily distinguish between the colour categories – no more than five is ideal, as in Figure 4.12.

- Pin maps allow us to pinpoint the exact location of things, such as social media check-ins, fly-tipping instances, or bomb sites. Pin maps are becoming increasingly popular as many data sources now include exact locations, such as from Twitter posts and Flickr photos.
- A series of regional map visualisations can often be more effective than one large global map visualisation. For example, when showing the locations of a university's alumni population, creating separate maps for cities and continents rather than cramming all the information into one large map will make the findings more visible and will likely drive social media engagement in these locations.
- Audience-appropriate stylings can be used for any type of map visualisation, such as illustrated cartoon maps for children or adapted standard maps in more formal settings. Additions such as arrows and speech bubbles – as in Figure 4.13 – can help your audience find conclusions quicker or add complementary information.

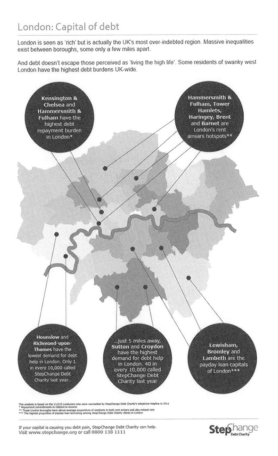

Figure 4.13 'London: Capital of debt' infographic

Credit: Produced by StepChange MoneyAware. www.flickr.com/photos/moneyaware/7164496820. CC BY 2.0.

Word clouds

- Word clouds or tag clouds are visualisations of text data. They work best for adding emphasis to experiences and opinions expressed in qualitative data. Words are usually displayed with the size of the font proportional to their importance or frequency in the data, and other design elements like colours and arrangement are often used to support this, as in Figure 4.14.
- Word clouds found popularity on the web through the 1990s and now social media and blogs have made them common occurrence again. They tend to be static image files that are quickly and easily created, with high engagement and click rates. Try Wordle.net as a starting point.
- In communicating research, word-based visualisations are often used effectively in linguistics and language studies, marketing studies and management studies. Political speeches are rich sources for word clouds, where, once viewed as a whole, the user can identify key themes in manifestoes (see Figure 4.15).

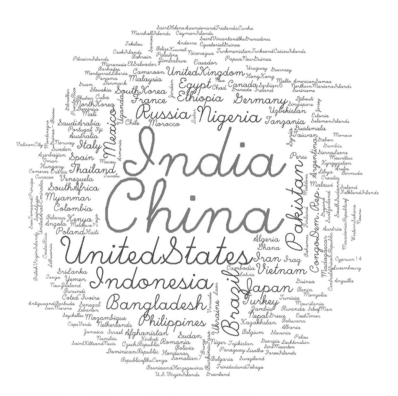

Figure 4.14 World population tag cloud created in R (programming language) with package word cloud

Note: Data is from 2010 and/or 2011, taken from list of countries by population. Note that the proportional sizes of China and India were divided in half.

Credit: Created by Seb951 (2011). CC-BY-SA 3.0 Unported. https://en.wikipedia.org/wiki/Tag_cloud#/media/File:Word_population_tagcloud_2011.png

- As part of a larger narrative infographic, you can use a word cloud to convey a single important number, percentage or word using impactful fonts and styling. Spelling out the message or key take-away can be more effective than using a chart, and it immediately signals to your reader what they should be focusing on. Viewers will read the text and then look at the chart to reaffirm what they read, not the other way around.

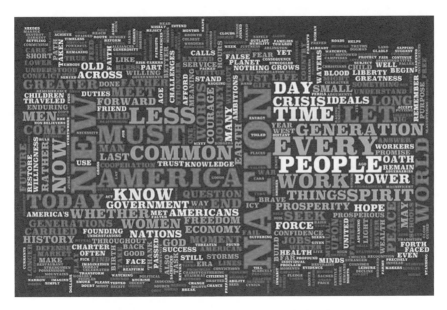

Figure 4.15 Barack Obama's inaugural speech as a word cloud

Credit: Created by Shawn Campbell (2009). Flicker, CC BY 2.0

Bar or column charts

- Bar charts are one of the best options for making easy comparisons between different discrete categories.
- In a bar chart, the bars are horizontally oriented, while in a column chart the bars are vertically oriented. A bar chart has rectangular bars of the same height with widths proportional to the values they represent, with each bar representing a particular category. Column charts are the reverse; the width of the rectangles is the same, while the height changes.
- Most multiple-choice survey answers – or anything that can be counted or categorised – can be visualised using a bar or column chart, such as in Figure 4.16 showing medal hauls from different countries.
- You can use bar charts to compare different groups or to track changes over time. Keep in mind, though, that bar charts work best when the changes being illustrated are larger and easier to see.
- Adding percentages or values to the top of each column further emphasises the results.

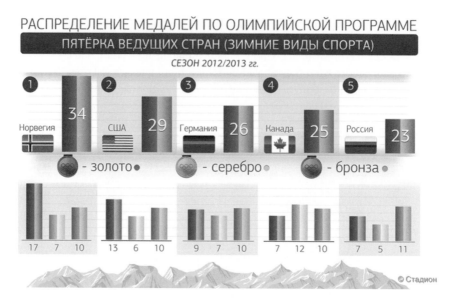

Figure 4.16 Results of pre-Olympic winter season 2012–2013 infographic, using columns to show gold, silver and bronze medal hauls

Credit: Stadium.ru (2013). CC-BY-SA 3.0 Unported. https://commons.wikimedia.org/wiki/File:Results_of_pre-Olympic_winter_season_2012-2013_infographic.png

Treemaps

- In a treemap, hierarchical data are displayed using nested rectangles which together represent a whole. Each branch of a tree is represented by a rectangle, which is then tiled with smaller rectangles representing sub-branches.
- When the colour and size dimensions are correlated in some way with the tree structure, the viewer can easily see patterns that would be difficult to spot in bar charts or area charts. They are also particularly useful for demonstrating which parts of the hierarchy have the largest or smallest quantities associated with them, such as in Figure 4.17.
- One of the main benefits of tree maps is that they make efficient use of compact space, so they can legibly display many items on the screen at the same time.
- Treemaps are becoming increasing popular in infographics, especially for analysis that requires more detailed views of a large amount of items. However, too many lines enclosing a small rectangle can make areas hard to read.

Line graphs

- A line chart displays information as a series of points connected by straight lines. The connections between the points allow us to visualise trends in data over time. If you have many data points, the line chart is the most effective chart to show these trends over time.

- Multiple lines can be shown on a single line chart, or a line can be laid over scattered data to show the best-fit trend. In most line charts, time is shown on the x-axis and measurements are shown on the y-axis.
- By displaying increases and decreases easily, line charts are more effective formats for showing overall patterns and trends in data than bar charts, as in Figure 4.18. However, line charts are not as good at showing specific values of data because the individual data points are harder to spot.
- It is sometimes hard to pick out the specific message of a line graph, so consider using a text description to strengthen the take-away message.

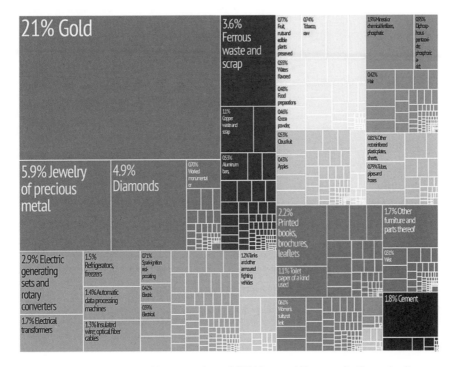

Figure 4.17 Lebanon Export Treemap from MIT Harvard Economic Complexity Observatory (2012)

Pie charts

- Circular charts divided into sections, pie charts are used to illustrate percentages. The whole circle represents 100 per cent, so each part of the pie stands for a portion of that 100 per cent.
- Pie charts are one of the most popular graphical charts used in the classroom, in the boardroom, in the news media and on blogs and social media.

- Pie charts are best for showing parts of a whole, especially when you have only two or three values, as in Figure 4.19. Too many values and it becomes hard for the reader to understand the story and grasp the angles and area represented, but adding labels and percentages as text can help to clarify.
- This format can easily be deceiving, if, for example, it is not clear what the whole stands for or if any of the parts are left out of the whole. Whether technically correct or otherwise, 3D pie charts are some of the worst offenders in bad visualisation.

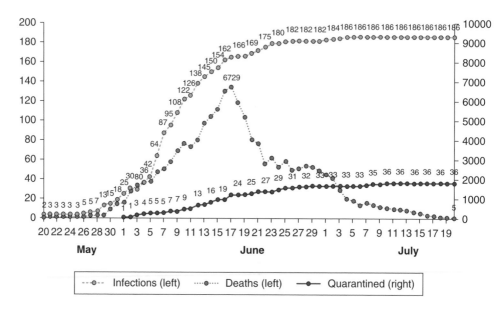

Figure 4.18   2015 MERS in South Korea data visualisation

Credit: Produced by Phoenix7777 using data from the Korean Ministry of Health and Welfare. CC-BY-SA 4.0 International. https://commons.wikimedia.org/wiki/File:2015_MERS_in_South_Korea.svg

Timelines

- Ideal for showing historical events, timelines are perfect for focusing on social, cultural or political change through the days, months, years and decades.
- Timelines use time series data to plot when events occurred along a central column, with horizontal lines as 'ticks' added for each time period, as in Figure 4.20.
- This format is much more interesting and engaging for the reader compared to using a table or list of dates. Creators can also add images, illustrations, text or pictograms to decorate the timeline, making it more shareable and memorable.
- Both interactive and static timelines are popular on blogs and social media.

Donald Trump's Hispanic Headache
% of Hispanics with a favorable/unfavorable opinion of presidential candidates

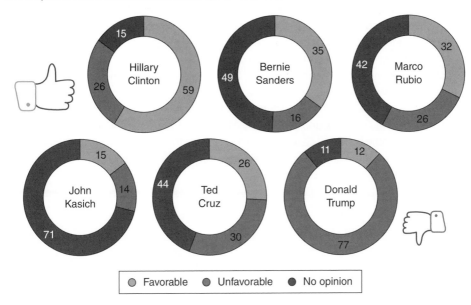

Favorable Unfavorable No opinion

Figure 4.19 Infographic featuring opinion data of presidential candidates, represented in pie charts/circle graphs

Credit: Created by Statista (2016). www.statista.com/chart/4523/donald-trumps-hispanic-headache

Figure 4.20 Part of the 'Women In Computing' Timeline

Credit: Created by Timeline JS. https://timeline.knightlab.com

Symbols and iconography

- Whichever graph, chart or map style you choose as the centrepiece of your data visualisation or infographic, you should also consider buttons, vectors, icons and other graphics to complement the data and enhance understanding. Symbols provide a great visual shortcut rather than having to type out a description.
- Consider how adding any combination of the following might enhance the message and flow of your visualisation: arrows, labels, stars, flags, logos, speech bubbles, numbers, percentages, dividers and buttons. Figure 4.21 shows a football-themed example.
- For the patient, creative and technically skilled, you can make your own symbols in Illustrator or other graphic design programs. But for the novice or for those looking for a quick answer, there are many free online symbol libraries that you can browse by keyword for ready-made symbols and icons.

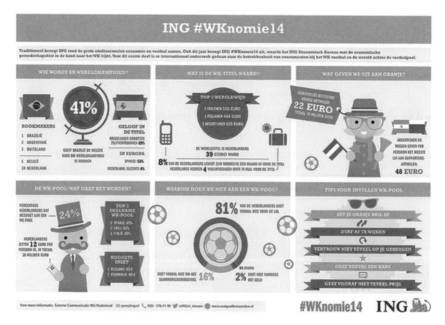

Figure 4.21   Football infographic

Credit: created by ING Nederland. www.flickr.com/photos/ingnl/14266948143 CC BY-SA 2.0

Where to find free symbols:

- The Noun Project – thenounproject.com
- Freepik – freepik.com
- Media Loot – medialoot.com
- FlatIcon – flaticon.com

This is by no means an exhaustive list of the options available to you when thinking about how to visualise your data and tell your story. Check out our suggested reading list at the end of this chapter or search for 'infographic + map + Brazil' – or whatever topic you're looking for – on Google or image-led social media platforms like Pinterest for more ideas. Try specialist blogs, data and methodology books, and flick back through the chapter to find inspiration from across the decades.

4.4.3   Tools for making your own infographics and data visualisations

The growth in the sharing of infographics and data visualisations on social media has sparked the creation of many free and inexpensive template-based tools and apps for novices. Some tools are aimed at complete beginners and allow for the simple creation of maps and charts; other tools have higher degrees of complexity and allow the user to create animated responsive visualisations. Most of these tools have easy upload, drag and drop features and give users the ability to input data from a spreadsheet, add text or logos and make some of the whizzier edits done by professionals, but with far easier functionality and with much more sophistication than a chart created in Word or Excel.

Every year, new online tools pop up, but below are some of our tried-and-tested favourites for creating infographics and data visualisations on a budget – or with no budget – starting with the easiest to use. Readers will note that we haven't included programs like Illustrator and Photoshop in our tools section. This is because we feel they can be quite cumbersome and time-consuming, and we love the quick social media sharing offered by the free online tools.

Free infographic makers including Infogram, Piktochart and Canva

These three tools are a great place to start and are similar in many ways: they provide ready-made, well designed templates and allow users to customise the look and feel of their work.

Infogram (https://infogr.am/) was built to help users create charts and infographics 'the easy way'. First, choose a template and colour scheme; then visualise the data by adding charts, maps, videos, images and icons; and finish up by publishing on social media through one-click share buttons. The Infogram team has won several awards, including winning Gold at the Information is Beautiful Awards – a coveted nod from some of the best in business. The range of charts and visualisations that users can create is fantastic: take your pick from bar, pie, line, bubble, area, multiple axes, tables, pictorials, countdown timers, gauges, word clouds and maps. On the data side, users can upload an Excel spreadsheet, a Google Drive spreadsheet or files from

Dropbox, OneDrive or JSON. Alternatively, you can browse their data library – which is connected to recommended open data sites like Eurostat and The World Bank – to search by keyword for datasets to visualise.

TOP TIPS FOR RESEARCHERS FROM INFOGRAM'S CEO, MIKKO JARVENPAA

1   **Do quick-and-dirty visualisations and test best ways to show your data**
    Infogram allows academics and researchers to make sense of large datasets. Brazilian think tank FGV/DAPP used Infogram to visualise complex data easily, shedding light on the dengue fever epidemic in their country. Once the data was visualised, they were able to pinpoint areas for further research. In their case, they spotted a region on a map with an alarmingly high amount of dengue cases that they had previously overlooked.

2   **Take advantage of the multiple data sources supported**
    Infogram makes it easy to upload your data in different formats, which is great for people dealing with a lot of important information from various sources. Our charts also update in real-time, which means changes made to your connected spreadsheets will automatically update in Infogram. People love this feature because it saves busy professionals time.

3   **Stick to the basics**
    Infogram offers over 30 different types of charts and visualisations. 80% of data visualisation needs can, however, be covered by good old line chart and bar/column charts. Also, as much as we all like the aesthetics of the round pie chart, it's often misused and does not add to the understanding of the presented data. There are special cases when the more adventurous visualisations like funnel charts or unit charts come in handy. They are good to have available in your special effects toolkit, but don't overuse them.

Piktochart (piktochart.com) is another easy-to-use infographic maker that allows users to create long, vertical infographics, reports, posters and presentations, with over 500 templates to work with. Piktochart is specifically geared to helping non-designers create stunning visuals, and helping educators and NGOs find ways to repurpose complex information. Users can edit text, fonts and colours, and add stock photos or upload their own. All the infographic templates are fresh and modern, with a wide range of themes. What we feel makes Piktochart special is its online Presentation Mode, which instantly presents your long, vertical infographic as if it were a slideshow. To help new users, Piktochart also has a well thought-through selection of free step-by-step guides, ebooks and tutorials.

Canva (canva.com) is the most accomplished of these three tools, 'designed to empower you to create incredible designs' for data visualisations and more, with a huge selection of templates for social media graphics, marketing materials as well as infographics. It is a more advanced tool, but for those interested in learning more about good designs and finding inspiration, its Design School online tutorials, blog posts and mailouts (https://designschool.canva.com) are fantastic resources. One-click shares and downloads are available here, making sharing now or saving for later very easy.

Free timeline makers including TimelineJS and HSTRY

If you're looking for a way to create static or interactive timelines, these tools offer easy input of data and multimedia content, allow users to customise the look and feel of their work, and are social media friendly. Again, there are more tools out there, but we recommend starting with these two.

TimelineJS (timeline.knightlab.com) is a very clever tool for making visually rich interactive timelines that can be embedded on your website or blog and shared through social media. It has been produced by the Northwestern University Knight Lab: a team of technologists and journalists working at advancing news media innovation through exploration and experimentation. By inputting your timeline data, adding accompanying descriptive text and pasting in links to content from Twitter, Flickr, YouTube, Vimeo, Vine, Dailymotion, Google Maps, Wikipedia, SoundCloud and more, TimelineJS pulls together a stunning customisable timeline. The result looks far more professional than a list of dates and descriptions – far more engaging for your social media audiences.

Zach Wise, TimelineJS's original creator and an Associate Professor at Northwestern University, told us:

> TimelineJS is an ideal tool for presenting various pieces of work in which chronology is important. In fact, one of TimelineJS's core capabilities is the ability to show the many events that lead up to a singular moment of insight or discovery – a capability that reflects the course that scientific or academic research often takes – and is one reason why the tool is useful for researchers. (Personal Correspondence, April 23 2016)

TimelineJS has been used by some big names. Wise tells us:

> In 2015 the *Daily Breeze* in California won a Pulitzer Prize for its work to uncover extraordinary executive pay in its local school district. To present the various pieces of journalism in one spot they used TimelineJS to great effect. Similarly, and in another instance that included Pulitzer-winning work, the *Boston Globe*'s Spotlight Team used TimelineJS to show readers key stories in its investigation into abuse by clergy in the Catholic Church. It's easy to imagine a researcher doing something similar with studies she's conducted or to highlight important work in her field in general. (Personal Correspondence, April 23 2016).

TOP TIPS FOR RESEARCHERS FROM TIMELINEJS CREATOR, ZACH WISE

1   **Find great material**

TimelineJS is designed to be easy to use and deploy, which allows you to focus on telling a story. You can tell a better story with great source material, so be sure to include interesting links, photos, videos, sound bites, and other pieces of media.

2   **Edit**

Users often include hundreds of entries in a single timeline, which leads to a less-than-ideal user experience. While thorough, this approach doesn't serve your reader very well. A better approach is to focus on telling an interesting story and to focus on critical, pivotal, or ground-breaking moments.

3   **Keep going**

TimelineJS is easily updated. Some instances of the tool have been updated consistently with new information for three years or more. So don't hesitate to add new events as they occur.

HSTRY (www.hstry.co) is a digital learning tool that enables teachers, students and historians to create and explore interactive timelines (see Figure 4.22). The team behind the tool created it to address what they believe to be inadequacies in the way history is often taught in the classroom: 'An answer for teachers' pleas for innovation in teaching history that captivates, tells stories, and speaks to students in the digital age we live in.' Users can attach text, images, video, audio and quiz questions, making it an interesting interactive timeline option for researchers, teachers and historians alike. Gamification of social media content is increasingly popular and often leads to high levels of engagement so even if you're not a historian the quiz element is a valuable feature.

HSTRY timelines can be shared through Google+, Facebook, Twitter, LinkedIn and Pinterest, as well as embedded on a website or blog. The HSTRY Community of educators will also be able to see your timeline, and likewise you can browse the work of others.

Free data visualisers including Google Fusion Tables and Tableau

Google Fusion Tables (google.com/fusiontables) is a data visualisation web application that gathers, visualises and shares data tables. It is a more complex tool than those listed above and is geared towards Google users with large datasets who already use Google Drive and Google Sheets, but it allows any user to create maps, charts and graphs. For datasets containing locations mapped with latitude and longitude points, Fusion Tables

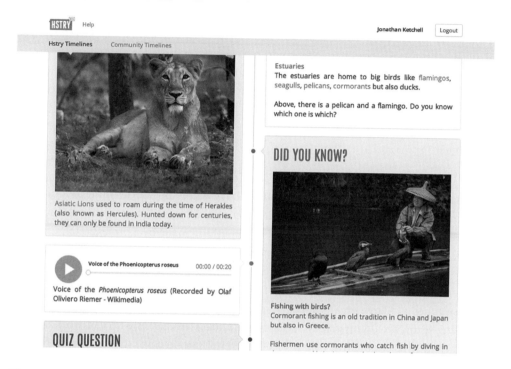

Asiatic Lions used to roam during the time of Herakles (also known as Hercules). Hunted down for centuries, they can only be found in India today.

Voice of the Phoenicopterus roseus   00:00 / 00:20

Voice of the *Phoenicopterus roseus* (Recorded by Olaf Oliviero Riemer - Wikimedia)

QUIZ QUESTION

Estuaries
The estuaries are home to big birds like flamingos, seagulls, pelicans, cormorants but also ducks.

Above, there is a pelican and a flamingo. Do you know which one is which?

DID YOU KNOW?

Fishing with birds?
Cormorant fishing is an old tradition in China and Japan but also in Greece.

Fishermen use cormorants who catch fish by diving in

Figure 4.22   Screenshot of the HSTRY application showing embedded audio, quiz question integration, item notes and other features

Credit: www.hstry.co/media-kit. CC-BY-4.0

auto-detects location data in a table and displays it as a customisable Google Map that can be embedded on a blog or website, or shared through social media.

What is special about Fusion Tables – just like the rest of Google Drive's functions – is the live team collaboration it enables. This is particularly useful if your project has many researchers who need access to a dataset at the same time. Team members can apply filters and download just the subset that's useful to them. Alternatively, the project lead has the ability to lock down the data and prevent its export.

Though focused on business customers, Tableau (tableau.com) is an option for any project looking for highly powered, reactive and interactive data visualisations. Once connected to your databases and spreadsheets, Tableau can generate a variety of professional graph types and customised dashboards, all shared online through the cloud. The charts have been 'algorithmically shaped and color-coded to conform to Edward Tufte-approved academic research – in other words, with no training at all, everyday employees can know their work is more or less in line with that of the most influential information designer of the 20th century', writes Mark Wilson on the *FastCoDesign* blog (2015). Tableau Desktop is free for students and teachers, but, as it is the most advanced of all the options explored here, we would recommend it more for experienced users.

If you're not into design-led infographics and find yourself more aligned with the Tufte school of thinking, these two tools above should definitely be first on your list to explore.

Hiring designers to create your data visualisations

Graphic design and visual data manipulation have whole businesses built around their creation, and you should accept that what you might be able to create using the tools we've discussed above may not be as slick as something made by a professional.

If visual creativity doesn't come easily to you or your skill level isn't very high, it's probably best to hire the skills of an experienced professional, or at the very least talk to some designers who can show you examples. If you are an academic attached to a university or a researcher at a large organisation, you may have access to the skills of a professional marketing team, including graphic designers and data visualisers who can work to your brief to create this content for your project. The brief should cover information about the audience you are trying to reach, the messages you want your infographics or data visualisation to convey, and the platforms through which the content will be shared. This is all important information for the designers, who will have a good understanding of the styles and approach more suitable for particular audiences. It's likely that the designers won't be able to give you much advice on what story to pick out of the data; their skills lie in the visual side of things, unless they have specialist backgrounds in science or geography, for example. You will need to go into the meeting with clear ideas and examples of design styles that you like or particularly don't like.

A real plus of working with professional designers at your university or institution is that they will make sure that the work you are commissioning fits in with the wider branding of the organisation. You will benefit from maximising the power of your institution's brand by having its logo featured in your infographic, for example, and from it being shared with a wide institutional social media following, including alumni and students. Plus, the professionals will have a clear idea of best practice, legal requirements and copyright issues, and can advise you of these.

The use of professional skills and guidance comes with a price tag, however, so explore calling in favours or ask any contacts for some pro bono advice, or make sure you portion off some of your budget or bid application to account for covering these activities.

4.4.4 Checklist: Are you ready to share?

Once you have made your first infographics and data visualisations, use this checklist to make sure you're ready to share your work.

Is it readable? Is the story you are telling in your work understandable? Can the reader understand the message, narrative or conclusion? Will they take away the right conclusion or is the story lost in messy design elements?

Are your data distorted? Have you used the most appropriate formats for displaying your data? Getting the balance right so that your design elements highlight your message and stimulate the viewer, on the one hand, and don't distract, deter or overwhelm, on the other hand, is not always easy. Accidental – or intentional – distortion of the data needs to be avoided at all costs.

Is the style appropriate? Colour palette, filters and fonts used in the infographics you're creating should all be uniform if they're part of a series or campaign. Think about the types of feelings you want to elicit from people and the associations with your topic.

Is it engaging? Is it appealing to look at? Is it attractive enough to catch the eye of a reader? Can you use a question format in the title to attract attention?

Is it correctly branded? Brand your infographics appropriately using your organisation or project logo, social media feeds, website URL and hashtag. This will help your audience recognise your content quickly when it appears in their feed. It doesn't need to be heavily branded – let your charts and maps be the focus.

Is it shareable? Is the infographic or visualisation in a format that can be easily shared through social media, like a jpeg, gif or mp4? Have you included the Twitter name of your project or a link to your website on the infographic? If you're embedding it on your blog, does your page include sharing buttons?

Is it correctly sized? You will need different sizes and dimensions for your files for different social media platforms. The optimum sizes change all the time and the exact specifications are easy to find online, but as a rough guide: for Instagram you'll be looking to share a square-shaped image; for Facebook, Twitter and LinkedIn you'll be looking to share a wide rectangle-shaped image; and for Pinterest you'll be looking to share a tall rectangle-shaped image. Some of the free online tools will automatically size and re-size for various social media platforms.

4.5   SHARING YOUR INFOGRAPHICS AND DATA VISUALISATIONS ON SOCIAL MEDIA

You've found a fascinating story to tell. You've planned your infographic with precision. You've spent time creating it and have the finished product ready to go. But how do you share it on social media in a way that will increase the likelihood of audience engagement, drive awareness of your wider project, and perhaps even help to make a policy or media impact? Having a social media following in some form is a start, but there are other things you should consider, which we'll now discuss.

4.5.1   Sharing infographics and data visualisations through blogs

Good infographics and data visualisations should be able to speak for themselves, but an accompanying blog post that explains where the data came from, the story behind the data visualisation, or extra details that enhance audience understanding can truly add to the value and shareability of your work. Plus, a blog post is a social media-friendly method of dissemination – with blog platforms built for one-click sharing and easy social share counting and, to add SEO weight, with keywords most likely scattered throughout your accompanying write-up. When it comes to sharing a blog post about your data visualisation through social media, you'll ultimately be sharing the URL to the blog post, which will give readers a base to come back to and read more about your work and your project; your visualisation isn't just floating out there in the world.

As discussed in Chapter 3, there are a few options with blogging: you could start a new blog for your project and all your data visualisations if you'd like to regularly publish and write about your work; or you could investigate university or institution blogs that are looking to publish research-based blog posts or maybe even specifically data visualisations; or you could look to wider industry blogs that specialise as knowledge exchange platforms for a community of readers interested in data visualisations. Some of the tools mentioned in the previous section have online data visualisation libraries too. Think about which of these platforms would be best for reaching your intended audiences and how the accompanying text for your visualisation can add value to your work.

Just like any blog post, your data visualisation blog post will need a title, and again there are a few options. A narrative title – for example, 'Infographic shows there is 50g more sugar in granola bars compared to muesli' – will likely produce strong search engine results for your infographic as it hits all the keywords a user might search for, and when shared on social media it quickly tells the story. Journalists, news hunters and general readers may find this style most appealing. A question title – for example, 'Do you know how much sugar is in your granola bar? This infographic compares sugar levels in break-fast bars' – does much more to encourage your readers to open the blog post and learn more: they need to actually open the blog post and look at your visualisation to learn the answer. This approach is more informal but follows best blog practice for providing a call to action. Alternatively, an exclamation title – for example, 'You won't believe how much sugar is in your granola bar! Be amazed by this infographic' – wins the award for the most Buzzfeed-ish millennial clickbait title, which may well get a lot of clicks, but it is unlikely to get seen by serious readers. Choose which one is most suitable based on your intended audience and the impacts you're working towards. If you're sharing on a selection of blogs, then that is the ideal time to experiment with titles and look to find out what works best using A/B testing.

If your data visualisation is a static image, you will likely be uploading a png or jpeg file (or sharing the html code for embedding the file), whereas a moving data visualisation could be a gif or mp4. Whichever file type you are uploading to your blog, add a clear description and fill out the metadata section to boost your work's SEO ranking. If you are tracking your non-academic citations with Altmetric Explorer or a similar tool, use the description to link to your DOI too.

Your blog platform may want to automatically resize your work down to fit the blog's page width but this may mean a less than optimum infographic viewing experience, so allow readers to click through to the full-size infographic. Uploading the full-size info-graphic is important for audience understanding; unless they are able to explore and see the full image in its entirety, the project and the content may well be lost. Lankow et al. (2012: 149) note that uploading the larger image as a click through also helps decrease the content's bounce rate (people visiting one page on your site and then closing the browser or hitting the back button). Additionally, if you are using Google Analytics, then having a separate trackable URL for the full infographic allows you to track exactly how many users go on to explore the full image.

4.5.2 Sharing infographics and data visualisations through social media channels

Twitter and Facebook are obvious choices as prime social media channels to share info-graphics and data visualisations based on academic research. Audiences here are generally looking for news and entertainment, have a few minutes spare to look at something in detail and are interested in sharing content that they connect with. Sharing the title to your data visualisation or a take-away finding along with your data visualisation file will help to attract more readers and engagements. Some examples of effective tweets or Facebook updates could include:

- New #infographic compares sugar levels in #granola bars: where is your favourite? + IMAGE FILE
- Gender #paygaps around the world compared in this #infographic – UK 19%, NZ 5%, Japan 26% + IMAGE FILE
- #Infographic: steady growth generates higher levels of #wellbeing among citizens than boom and bust cycles + IMAGE FILE

Image-based social media platforms and apps like Instagram and Pinterest are also certainly worth considering (especially Pinterest for its active communities around teaching and learning resources), but be sure to spend time learning how individuals and institutional accounts interact in this space and which accounts might be worth reaching out to directly. Eye-catching data visualisations from a variety of sources have been re-pinned tens of thousands of times on Pinterest, as the platform has found a growing user base of designers and data visualisation enthusiasts hungry for inspiration. Some of the larger data visualisation accounts on Pinterest have group boards, where all followers are invited to pin their images. They look especially good on Pinterest because of the site's vertical layout. As well as sharing your own, consider re-pinning the infographics of others so long as they are relevant, interesting and correct.

Video-driven platforms such as Snapchat and Instagram are the place to share data visualisations aimed at younger audience segments. For children and teenagers who don't use Facebook because their parents' activities are just too embarrassing to endure, Snapchat's quickly disappearing photos and videos have seen the platform grow to 150 million users, the majority of whom are between 13 and 24. A similar audience on Instagram has also been attracted to the exclusivity of a no-grownups platform, but here the content published is mostly permanent. Outside of these audience groups, Twitter and Facebook have learned from the popularity of video elsewhere and recently got on board with video in a big way, with autoplay functionality for gifs and mp4s now a central part of these platforms' engagement strategies.

Data visualisation is a topic that garners lots of interest among developers and computer scientists and there are very active communities on Reddit and Github forums if you are looking to engage a more data-friendly audience. See, for example, the DataIsBeautiful Reddit discussion (reddit.com/r/dataisbeautiful), which had in August 2016 over 7 million readers. But we do stress *engage* here, as self-promotion on Reddit is extremely frowned upon. It is always a good idea to familiarise yourself with a community before sharing. Another option for Reddit engagement is to get in touch with moderators about taking part in an Ask Me Anything live interview where you can talk more widely about your research and the visualisations you've created (see McCloskey (2016) 'Using Reddit as a tool for public engagement, profile raising and scholarly dissemination' on the *LSE Impact Blog* for more on this).

But there are many other new and emerging social media platforms out there too. It is important to investigate your own individual and organisational networks across all social media platforms, identify key influencers and look to see what partnerships can be developed to help share your message. If you rarely use Twitter or Facebook, then it is unlikely that randomly sharing a captivating infographic on these platforms will get much notice.

For those who already have a social media presence, put the effort into building your credibility as a useful source of infographics and social media users will recognise this and return the social favour by sharing your work. Most communication, social media notwithstanding, is built on trust: if people trust the content you are sharing and the sources you are drawing from, they will be more willing to share on your behalf. Consider as a starting point the groups and networks you are already active in and look to grow from there. LinkedIn and Facebook can be particularly worthwhile spaces for close-knit, industry-specific interactions.

Hashtags across all these platforms are also an effective way to reach beyond your individual network. However, be sure the hashtag is active enough to reach people but not so populated that your content will get lost in the noise. Consider using both general hashtags and hashtags specific to the topic of your work – for example, general hashtags that browsing users might search for would be #dataviz #infographics and #datavisualisation, and more specific tags that users undertaking a more targeted search might use could be #history #womenshistory #genderpaygap and #feministdataviz, for example.

Perhaps the most important element of sharing on social media is learning what works. Measuring your efforts and keeping a record of the analytics, even if it is as basic as how many retweets or likes posts you are receiving, will help you develop and refine your social strategy. Not only will this recording help you understand the reach and potential impact of your current work, it is also incredibly useful knowledge for your future projects and can save you time and stress down the line when it comes to facing these questions again.

4.5.3   Sharing infographics and data visualisations in the 'real world'

After you've put all the hard work into creating your infographics or data visualisations, be prepared to discuss them at every opportunity! Conference presentations, public lectures and festivals are obvious venues to tell an engaging story about your data using your visualisations, but there are plenty of other spaces. Many universities organise informal Show-and-Tells, TEDx talks and Fringe-style events aimed at public audiences. Consider where your findings might provoke discussion and debate and look to see if there are any opportunities for engaging these audiences in dialogue. And be open to criticism! If the conclusions aren't as self-evident as you think they are, then that is incredibly useful feedback to receive and should inform your communication efforts.

4.5.4 Sharing infographics and data visualisations through targeted emails

Email marketing allows you to speak directly to your audiences, segment those audiences easily and send multiple messages as part of a campaign. Though not as sexy as social media, email is still a considerable driver of web traffic and this is even more pronounced among mobile internet users. Figures from Campaign Monitor suggest that you are six times more likely to get a click through from an email campaign than you are from a tweet (Beashel, 2014) and for businesses email marketing has an ROI of 3800% (Direct Marketing Association, 2015). According to 2014 research by RadiumOne, 'Globally, 82 percent of content shared on mobile is shared through messaging, email or text' (Southern, 2016). Infographics and data visualisations are unlikely to be exceptions to this, although, by their nature, 'dark social' patterns are incredibly difficult to measure. This means that it can be difficult to fully understand the extent to which your content is being shared. But it also means you should feel free to take advantage of the medium most frequently used for sharing content.

To make the most of this channel for sharing infographics and data visualisations, use a newsletter tool such as MailChimp or Campaign Monitor to create a design where your data visualisation is the first thing your readers see when they open the email. Use the email subject line to add a call to action and make opening the email an irresistible proposition. As with a blog post, add a paragraph to briefly highlight some of the key messages and include links back to your project blog post or website, social media channels and any full report or findings that built the infographic. You could use standard GMail or Outlook email software but using a more professional tool will allow you to track the number of opens, clicks and other engagements within your emails – useful information to know for future mailouts or project reports. Overall, this is a fairly simple yet effective way of getting your content seen, and will complement your social media strategy nicely.

When it comes to building your email lists, add and segment colleagues past and present, research contacts you've met through networking and journalist contacts if you're seeking media coverage. Segmenting the lists means you can send separate emails and invitations to these groups and tweak the messaging and tone accordingly. A quick note on building subscriber lists: do not spam 'cold' contacts that you've never met or interacted with, but ensure that your contacts would instead be happy to hear from you. And don't forget to share on relevant mailing listservs.

Whatever method you pursue, make sure to coordinate a plan for sharing your data visualisations and infographics. Focus on what specific audiences will be most interested in and will benefit the most from your data and look to build attention from there. In most cases, an audience does not happen overnight. Continue to pursue opportunities where conversations, both online and offline, can occur and ensure you make notes on what works for the future. And, if you are looking for further guidance, Chapter 7 includes more detail on choosing the right platform and planning a coherent social media strategy.

4.6   CONSIDERATIONS

Finishing up our focus on infographics and data visualisations, here are a few final points to consider before you get started:

- **Distortion of data**: Getting the balance right so that your design elements highlight your message and stimulate the viewer, on the one hand, and don't distract, deter or overwhelm, on the other hand, is not always easy. If in doubt, before publishing and sharing with the world, do some user testing and ensure that your message can still be understood clearly – and that it's the right message. Accidental – or intentional – distortion of the data needs to be avoided at all costs. Your viewers will value accurate information and will share it only if they trust it, so build a good reputation for your project by sharing good, clean data in a comprehensible and clear style.
- **Creative commons**: Many online research projects publish their work with an attached Creative Commons licence. With an attribution licence, images and infographics can be re-used by anyone, for any purpose, providing credit is given to you as the producer of the image. Deciding whether to allow your images to be used under an Attribution licence is a personal decision and should be given careful thought, as granting unrestricted permission will allow profit-making outlets or political groups to use your images. Some institutions and employers will recognise the concerns of their researchers and staff, and may advise against this form of licence. However, as noted in UCL's 'Communicate your research' online resources for academics in the Mathematical and Physical Sciences, placing restrictions on your images is 'a guaranteed way to reduce the public impact of your work' (UCL, 2016a).
- **Underlying data**: Speaking of impact, a primary way to ensure your infographic is re-used, built upon and remixed is if you make sure the underlying data and code for your research are available for others. At the very least (and with the complications that exist for human-subject data), make sure your data are properly archived and preserved for future reference.

4.7   FURTHER READING

D'Ignazio, C. (2015) 'What would feminist data visualisation look like?' [Blog post]. *MIT Center for Civic Media*. December 20. Available at: https://civic.mit.edu/feminist-data-visualization

FiveThirtyEight podcast series 'What's The Point'. Available at: http://fivethirtyeight.com/tag/whats-the-point
Explores stories on how data are changing our lives.

Kimball, M.A. (2006) 'London through rose-coloured graphics: Visual rhetoric and information graphic design in Charles Booth's maps of London poverty.' *Journal of Technical Writing and Communication*, 36(4): 353–381.

Lankow, J., J. Ritchie and R. Crooks (2012) *Infographics: The Power of Visual Storytelling.* Hoboken, NJ: Wiley.
Full of examples and ideas for the novice or the experienced infographic creator.

Lupi, G. and Posavec, S. *Dear Data* postcard project. Available at: www.dear-data.com

McCandless, D. (2009) *Information is Beautiful*. Frome: Collins.
Stunning displays of information on social trends, nature, war and more.

Piktochart's free ebooks and resources, including 'Create your first infographic in 15 minutes'. Piktochart ebook. Available free at: http://piktochart.com/wp-content/uploads/2015/07/Piktochart-e-book-2-Create-Your-First-Infographic-In-15-Minutes.pdf

Tufte, E.R. (2001) *The Visual Display of Quantitative Information*. Second Edition. Cheshire, CT: Graphics Press.
Edward R. Tufte's classic look at good practice for visualising data.

5

CREATING AND SHARING AUDIO AND PODCASTS

The radio would be the finest possible communication apparatus in public life, a vast network of pipes. That is to say, it would be if it knew how to receive as well as to transmit, how to let the listener speak as well as hear, how to bring him into a relationship not isolating him. On this principle the radio should step out of the supply business and organise its listeners as suppliers.

– Brecht (1932: 15)

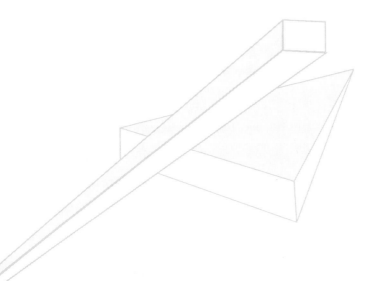

In the Autumn of 2014, a new podcast series was launched which would go on to forever change the way the public in North America, the UK and wider would interact with and understand the potential power of on-demand audio. *Serial*, a 12-part spin-off from the popular radio and podcast, *This American Life*, kicked off its inaugural season with an investigation into the murder of a Baltimore high school student, Hae Min Lee. It was an automatic hit. The series was characterised by its candid conversations between Sarah Koenig, the podcast's executive producer and host, and Adnan Syed, Hae's ex-boyfriend, who was accused and convicted of her murder in 2000. Koenig and her team of producers delved deep into the details surrounding Syed's conviction, unearthing other suspects and alternative theories along the way. It was a classic whodunnit that ignited the interest of global audiences like no other podcast before it. *Serial* listening parties sprang up around the US and the UK, along with lengthy Reddit threads devoted to solving the case, and – in a meta way that can only be spawned by the fog of an internet frenzy – it even spawned podcasts about the podcast.

Before *Serial*, the tradition of on-demand and on-the-go audio was already rich with medium-defining examples, such as *Radiolab* and *This American Life*. Radiolab popularised topics in science through its narrative-heavy style and unique sound textures, whereas *Serial*'s forebearer, *This American Life*, a long-running radio show turned podcast, hosted by Ira Glass, likewise captivated audiences with its distinctive brand of storytelling – a three-act show centred around themes like 'Babysitting' and 'Scenes from a Mall'. *Serial*, however, managed to bring previously unengaged audiences to the medium quicker than any previous show. Because there wasn't a precedent for this sort of fervour around podcasts, no one anticipated the level of the show's success from the outset, not even Koenig. Before it was released, she was expecting, at most, 300,000 downloads (Daniel, 2015) which was a fair and ambitious estimate for the medium at the time. 'I never meant to create a fever,' said Koenig to a group at Penn State University, 'It's hard for me to answer why [it went viral]. I didn't know if it was even going to work' (Berry, 2015: 170). *Serial* had reached 5 million downloads on iTunes faster than any podcast ever; and, as a whole series, it was downloaded over 78 million times by mid-2015 (that's 6.5 million downloads per episode) and created a wide-reaching excitement previously unknown to the podcasting world (Daniel, 2015). And although it was not as loved by critics as its inaugural season, *Serial*'s second season, about the disappearance of an American soldier from his army base in Afghanistan, still surpassed 50 million downloads (Robinson, 2016).

A poll of *Serial*'s listeners undertaken by the creative ad agency McKinney, showed that nearly a quarter of the podcast's listeners had never heard a podcast before, with nearly half going on to listen to podcasts on a weekly basis as a direct result of the show (PR Newswire, 2015). From the same poll, a staggering 90 per cent of those first-time listeners said it changed the way they thought about podcasts. This was all part of the so-called '*Serial* Effect' (PR Newswire, 2015). Comedy podcast *WTF with Marc Maron*, most known for his interview with Barack Obama (in Maron's garage, no less), saw monthly downloads increase to 5–6 million a month after *Serial*'s premiere (Mallenbaum, 2015). The podcast series *Criminal*, a show about people-related crimes from geological theft to murder, also saw a change, with host Phoebe Judge noting that there was 'a whole new audience coming to podcasts that appreciated *Serial* and the quality of that show' (Mallenbaum, 2015).

And so, at last, the podcast was having its moment to shine in the glory of the spotlight, nearly a decade after the first of its kind came on the scene out of a university research centre (more on this later). But, of course, a phenomenon like this doesn't exist in a vacuum and *Serial* alone can't account for the totality of the medium's resurgence. The ubiquity of smartphones has been the biggest boon for podcasts in recent years. Additionally, podcasts are an increasing feature on car journeys through Bluetooth technology (Zorn, 2014). Shortly after their launch in the early 2000s, podcasts did have a moment of popularity and buzz, but this faded because of the cumbersome nature of downloading a podcast: you first had to download one on to a computer and then transfer it to a portable media player. But the launch of the iPhone in 2007, which allowed a listener to download it straight from WiFi on to their phone, put them back on the map (Kang, 2014).

In this chapter, we argue that researchers, students and academics working in all fields can take advantage of this podcasting renaissance in order to reach wider audiences and share work from multiple parts of the Research Lifecycle. Creating a production-intensive podcast like *Serial* may be unrealistic for budget-stretched and time-poor research professionals, but there is a range of podcast forms that do lend themselves to busy schedules and the tightest of budgets.

First, we will explore what podcasting is and where it came from before moving on to the benefits the medium can afford you as a research professional and knowledge worker. Then we will move on to explore the different types of podcast format, and then how to get started creating your own in order to share your research or project at any stage of the Research Lifecycle. We will also be looking at how best to share your podcasts using social media platforms appropriate for your audiences, and how to avoid copyright and licensing problems with your own content and that from any contributors. Whether you are a young researcher looking to reach new audiences with your ideas, a retired professor considering using audio technologies to bring new life to your PhD fieldwork from decades ago, or an NGO or charity communications professional looking for something innovative for your team to create, this chapter will set you up with all the tools and knowledge needed to launch a podcast for your project.

5.1   DEFINING PODCASTS

The term 'podcast' was first coined by journalist and broadcaster Ben Hammersley in his 2004 article for *The Guardian* on how traditional broadcasting was starting to face disruption from a new subset of amateur broadcasters (Hammersley, 2004). Hammersley, musing over how we should define the new format at the heart of the boom in amateur radio content, asked 'But what to call it? Audioblogging? Podcasting? GuerillaMedia?' Ever since he posed the question, 'podcasting' as a term stuck, and by 2006 it was indeed so commonplace as to be named 'Word of the Year' by the Oxford American Dictionary. The Dictionary now defines it as 'a digital audio file made available on the Internet for downloading to a computer or portable media player'. In practice, that means podcasts are downloadable audio shows accessed through platforms like iTunes or 'podcatching' apps, and played on computers, smartphones or MP3 players.

The advent of automatically updating feeds, or RSS feeds, laid the groundwork for the first ever podcast. In 1999, web browser company Netscape launched the RSS feed as a means to allow users to smoothly aggregate continually updating text-based content on the web, such as news articles and blog posts. In March of that year, Dave Winer, a software developer and blogger, adapted these feeds to include audio for the first time (Quirk, 2015). Winer would later join The Berkman Center for Internet and Society at Harvard Law School, where he met radio show host Christopher Lydon. Winer created a special RSS feed for Lydon, who then used it to upload audio files of political discussions and commentaries to his blog at around the time of the American invasion of Iraq. The audio blogs eventually became the *Radio Open Source* podcast, which is still running today. Lydon hoped that he could use this new, untethered medium to challenge the status quo and engage audiences in important debates about power, terror and democracy, a role he felt traditional media at the time had failed to fulfil when it came to questioning George Bush's invasion. Describing the situation, Lydon felt there was 'a vacuum of intelligent discussion. A complete vacuum. The lights were going out in American public conversation and democracy. My whole drive was: how do we fight this nonsense?' (Walker, 2015). For Lydon, here was a new medium that was revolutionary and anti-establishment, in the same way that blogs were: 'Anybody can be a podcaster. You can record your songs, your deepest thoughts, your silliest thoughts, your sonnets that you've been writing for years – you can spread it to the world' (Walsh, 2011).

5.2 A RECENT HISTORY OF PODCASTS

5.2.1 Podcasts go mainstream

Further and concurrent web innovations were key to launching podcasts into the mainstream in the US. In 2005, Apple added podcasts to the iTunes Music Store and, at the same time, podcasting directories, networks and new tools for making podcasts sprang up – including Odeo, which would later change direction to become Twitter (Quirk, 2015). These technologies meant podcasts could fit practically into busy lives on the go and they became a relative overnight success. In 2004, a Google search for 'podcast' would have yielded merely 6,000 hits. By 2005, it would come to yield 61 million hits (Berry, 2005: 144). According to Google Trends, searches for the term 'podcast' have remained consistent over time, peaking again in December 2014, the same year *Serial* premiered (see Figure 5.1).

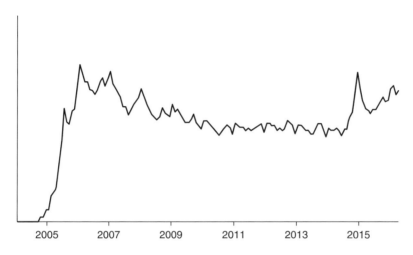

| 2005 | 2007 | 2009 | 2011 | 2013 | 2015 |

Figure 5.1   Google search trend for the word 'podcast'

Credit: www.google.com/trends/explore?date=all&q=podcasts. April 2016

Podcasting today benefits not just from the *Serial* Effect and the increase in smartphone usage, but also from newly evolving ventures that are redefining tastes and consumption patterns. The American collective *Radiotopia* – which succeeded in creating the most successful Kickstarter campaign in the radio and podcast category to date – publish a range of podcasts from the informative to the experimental. Other companies and collectives, such as Gimlet, Panoply (part of the online magazine, *Slate*) and Midroll Media, are succeeding in capturing high audience numbers even amid a flood of content (Quah, 2015a). In 2011, as the *New York Times* and the *Boston Globe* were turning away from podcasts, *Slate* persisted, so sure was it that the increasing take-up of smartphones would only increase the audience (Phelps, 2012). And that has paid off for the media company. *Slate*'s Panoply podcast audience tripled from 2014 to 2015, reaching nearly 6 million downloads a month (Owens, 2015).

Slate's podcast also increased in length in those years. Chief Content Officer, Andy Bowers, said of the increase in length: 'Where else on the Internet do people go and seek out something that's 45 minutes long and listen to it week after week after week, year after year? That kind of engagement is pretty extraordinary' (Phelps, 2012).

It is that engagement that makes podcasting for big media companies a surprisingly profitable enterprise. Advertising slots for shows with such an ardent fan base come at a premium. Gimlet survives on revenue from ads creatively placed at the beginning, midpoints and end of shows. *Slate* also survives on this advertising-based business model (Owens, 2015; Wang, 2015).

With podcasting so commonplace now, it is hard to imagine how innovative the first ones back in the mid-2000s were, but this new means of recording and distributing audio content ushered in a new era of broadcasting. Just as in the case of the blogging vanguards, both podcasting amateurs and the new wave of pros were able to bypass the usual media gatekeepers: anyone could get in front of a microphone and transmit their thoughts without worrying about advertising revenue or regulatory constraints. It is this very off-the-cuff style of early podcast venture and the possibility to subvert traditional media power structures that attracted podcast innovator David Winer to the medium: 'I felt that a podcast was like a blog in that it was something amateurs would do, not professionals … it would open up media to all kinds of people that it wasn't open to before' (Walker, 2015).

Winer's prediction that amateurs would dominate the medium didn't hold up entirely. Seeing the advantage of podcasts for growing their audiences, conventional broadcasters started offering their radio programmes as podcasts early on. After witnessing some successes with its first podcasted show, *In Our Time*, the BBC quickly grew its audio offerings to 20 in just two years, taking in breakfast radio shows, specialist music shows and comedy and drama programming (Berry, 2005: 150–151). The BBC World Service followed by podcasting its content, beginning in 2007. In just eight years the international broadcaster reached 1 billion podcast downloads. National Public Radio in the US, the Australian Broadcasting Corporation and the Canadian Broadcasting Corporation also followed suit (Berry, 2005: 150–151).

5.2.2   What drives the popularity of podcasts?

Podcasts may be liberating tools for broadcasters and amateur content makers, but they are also liberating for listeners, who, with the advent of iTunes and iPods, are now untethered from terrestrial broadcast schedules. Choosing the selection of shows they want to listen to and when, listeners become both curators and audience members (McClung and Johnson, 2010: 91). A study conducted in 2008 examined the motivations behind podcast listening and found that podcasts are primarily popular because of their entertainment value and quite simply because they made listeners 'feel happy' (2010: 88). Second, people downloaded and tuned -in because of time-shifting, meaning they could listen whenever and wherever they wanted and to whatever they wanted to. Third, listeners liked that they could build a virtual library of episodes. Podcast listeners also enjoyed the social aspect of content, the fact that podcasts spurred on conversations with friends and acquaintances. In fact, the social aspect of podcast listening was the most important predictor for podcast use (2010: 88). Additionally, podcasting is an intimate medium and one that spurs creativity, akin to reading. The fact that there are no visuals in podcasting may work to its advantage. Jad Abumarad, co-host and producer of *Radiolab*, has said of podcasting, 'In a sense, I'm painting something but I'm not holding the paintbrush. You are. So it's this deep act of co-authorship, and in that is some potential for empathy' (Weiner, 2014). Much like a novel as opposed to a movie, podcasts, when done well, excite the imagination and allow the listener to place themselves more easily in another world (Fitzgerald, 2015).

Similarly, the very medium of podcasting may have brought back the engrossed and captive audiences of radio's past before the invention of the television. In observing what attracted fans of *This American Life* to the show, and more specifically what allows them to 'visualise' the stories and build empathy for the people who feature on the non-fiction documentary programme, researcher Kristine Johnson found that audience curation and time-shifting helped to facilitate engaged listening. In other words, the fact that the audio had been pre-selected and could be consumed wherever was convenient for the listener, meant that listeners were more likely to be fully engrossed in the content than to simply treat it as 'background' fodder (Johnson, 2007).

In examining the appeal for podcast producers, Kris Markman and Caroline Sawyer offer a detailed look at the motivations behind independent podcasters (Markman, 2012: 547–565; Markman and Sawyer, 2014: 20–35), finding that many had become interested in the medium to reach niche audiences (Markman and Sawyer, 2014: 31). Others were motivated by the way podcasting gave them a platform for creativity and self-expression (2014: 32).

5.2.3 Academic soundscapes

The podcasts we published with the Public Policy Group (PPG) at the London School of Economics supported and diversified content from a range of our academic blogs. Our blogging team was enticed by the low entry investment, but most of all we were energised by the possibilities of the medium. As we found out, a recorder and mic can bring you into unexpected places. In one episode for the *LSE Review of Books* podcast, our host Amy Mollett and producer Cheryl Brumley took a walking tour through Chinatown in central London with two former Middlesex University academics, Professor Rosemary Sales and researcher Xia Lin. Sales, Lin and a team of researchers studied the sociology of the migrant Chinese communities living and working in the area. In the podcast, we wandered in and out of traditional meeting places, and spoke to an organiser at the local community centre while Sales and Lin introduced their findings on diasporic identity. The podcast also had an accompanying map in which additional sound and photo slideshows were embedded. The goal was to bring audiences into the studied space through sounds (to listen, go to: https://soundcloud.com/lsepodcasts/lse-review-of-books-episode-6).

Instead of just reading about the way Chinese communities in London use Chinatown, we wanted listeners to hear from people directly. An organiser in the Chinese community centre talked to us about the kind of immigrant groups that attend classes at the centre. A woman busily chatting to her friends at the pagoda, the traditional meeting place, told us that she comes to Chinatown weekly to do a shop and to attend a church nearby. In addition to the interviews, the 'wildtrack' (a term used in some radio communities to denote running background noise) on the Chinatown podcast provided a rich texture of sound from the area, including street noises, exercise class tapes and food trucks making their morning deliveries.

The podcast garnered 1,600 listens on Soundcloud and another 1,000 on iTunes. It served as an early example to our team about the value podcasting could bring to a piece of research, that by mixing in the monologue and dialogues of the researchers with a host of other sounds and encounters, we were also allowing listeners another way into understanding a complex subject (in this case, diaspora and identity).

In her chapter on how librarians can use podcasting, Professor of Creative Media Tara Brabazon also provides a means by which to understand how sound can be used in academia more generally to allow publics to grasp difficult subjects:

> Sound is a mode of communication that slows the interpretation of words and ideas, heightens awareness of an environment and encourages quiet interiority. It punctuates buildings, workplaces, leisure complexes and family life. The visual bias in theories of truth and authenticity means that sounds are often decentred or silenced in empowered knowledge systems. (2012: 140)

As a result, she argues, sound is rarely given a chance in academia to live up to being used to its full potential:

> Education rarely manages this sonic sophistication ... Formal educational structures are geared to develop literacies in managing print. Too often, soundscapes are cheapened with monotone verbal deliveries in lectures, interjected with stammering and confusion, and do not open our ears to other rhythms, melodies, intonations and textures in the sonic palette. (2012: 140)

Brabazon is really arguing for a more professionalised approach to the use of sound in academia, 'standards' over 'standardisation' as she calls it (2012: 140). But another lesson can be reaped from her passage: since academia favours visual interpretation over other senses, print over podcast, sound's role in helping us understand the world around us, has been sorely diminished.

The LSE Public Policy Group podcasts were first an experiment but then became an invaluable means by which we allowed new audiences to interpret events and phenomena; an aural complement to the very visual PPG blogs.

We repeated the sociological soundscaping in a podcast on the London riots for the *LSE British Politics and Policy* blog. Les Back, a sociologist at Goldsmiths University of London, who is well known for writing about sound, notably in his co-edited volume *The Auditory Culture Reader* (Back and Bull, 2003), brought our host and our producer through the neighbourhood of Catford in South East London, where riots took place in the summer of 2011 (to listen, go to: https://soundcloud.com/lsepodcasts/sets/british-politicast).

Back recalled the night's events and at the same time discussed the sociological under-pinnings that might account for the seemingly spontaneous riots. The podcast was also published with an accompanying map and blog post. Back first started the walking tour by bringing us to common areas for police-enforced stop and search, and then onwards to the shopping park where the looting and destruction took place. Back lived only minutes from the scene of the riots but had seen the night's events unfold on his computer as young people uploaded videos on to YouTube. He ends by saying that both the left and right of society have had the 'incapacity to hear and to engage with young people'. All the while, wildtrack of street scenes, including sirens and people chatting hurriedly on their week-day shop, were playing in the background. Both the Chinatown podcast and the Catford riots podcast drew in audiences of around 5,000 each through Soundcloud, iTunes and the *LSE Review of Books* and *LSE British Politics and Policy* blog. Compared to podcast mega-hits like *Serial*, which bring in millions of listeners and downloads, 5,000 seems modest. However, compared to journal article downloads, which have a very modest reach, usually numbering in just the hundreds, these podcasts, and a host of others produced by our team, reached a significant amount of people on platforms not usually associated with academia. In the end, what we hope these podcasts served to highlight for these audiences is that research and academic concepts can be presented in a variety of forms which don't necessarily conform to the conventions of your typical academic discourse. We tried to merge different media with academic research not only to make it more accessible to wider publics, but also to diversify the means by which we were trying to disseminate research.

Les Back, with his co-author Michael Bull, a reader in media and film studies at the University of Sussex, argue in *The Auditory Culture Reader* (2003) that academics and researchers need to stop privileging visual culture at the expense of our other senses. It is a sentiment shared by Brabazon (2012). Specifically referring to our auditory sense, Back and Bull write that 'thinking with our ears offers an opportunity to augment our critical imaginations, to comprehend our world and our encounters with it according to multi-ple registers of feeling' (2003: 2). Johnson's aforementioned study on *This American Life*, based on a survey of listeners, found that audiences of the show responded to the effec-tive and enthralling storytelling, which harkens back to the 'Golden Age' of radio (i.e. the 1920s through to the 1950s) (Johnson, 2007).

We can therefore think of podcasts as the new version of the old and proven medium of radio. Radio has filled our living rooms with aurally rich news reports and features for over a century. Ever since then, print was no longer the only means by which we came to understand events and phenomena outside our immediate purview. We had sound to transport us (and later, as discussed in the next chapter, moving pictures). Academics and knowledge workers hoping to harness the power of audio are operating in a great egalitarian age where mere novices have the platform and the means to broadcast their research beyond the strictures of journal articles and audiences likewise have more freedom to discover content, no longer tied to terrestrial radio schedules.

5.3 WHY PODCASTING IS USEFUL ACROSS THE RESEARCH LIFECYCLE

Although the transmission of sound is an old idea, its repackaging into a podcast format has transformed it into a surprisingly versatile medium. It is also a medium which lends itself naturally to the dissemination of academic or research material. Course administrators at Duke University, North Carolina, US, discovered this early on in 2004, when the university equipped incoming students with iPods preloaded with lectures (O'Brien, 2006). Duke was an early innovator in educational and instructional technology. Before their iPod initiative, they experimented with ways to incorporate things like laptops and Palm Pilots into the classroom. 'The Duke First-Year Experience', as the iPod project came to be known, was an extension of the university's goal to find new uses for technologies in pedagogy and was carried out, in part, with the help of Apple. The Belkin Corporation attached microphones to the iPods, so students could also submit their own content for evaluation. This was especially helpful in language courses where students would hand-in audio files throughout the semester to keep track of their progress. Faculty who incorporated iPods into their class said it led to an increase in student engagement (O'Brien, 2006).

In Duke's evaluative report a year on from the initiative, they claimed that the Belkin recording feature was actually the most beneficial feature, with 60 per cent of first years claiming they used it for academic purposes (Brabazon, 2012: 142).

Other universities followed in Duke's footsteps, which paved the way for Apple's iTunes U – a virtual classroom with podcast lectures from universities around the world (Eckstein, 2013: 51). In 2007, Apple launched the platform in conjunction with universities. It featured free lectures, events podcasts and campus orientation tours. Courses from big-name institutions like the Massachusetts Institute of Technology (MIT) – a pioneer in Massive Online Open Courseware (MOOCs) – the University of Michigan and Stanford University quickly grew in popularity. Seeing the possibility for the platform to aid learning in the classroom, Apple later re-launched iTunes U with a greater focus on course management akin to Moodle (Fenton, n.d.). Early podcasts on iTunes U rarely ventured away from lectures, which still comprise the majority of educational podcasting content (Brabazon, 2012: 143).

A number of non-university-based research institutes and not-for-profits have also entered the podcasting space. To name a few, the UK education not-for-profit Jisc has a podcast series on digital technologies in Higher Education and research; the Royal Society of Chemistry has a series on elements and a separate podcast featuring interviews with science writers; the Wellcome Trust publishes its lunchbreak lectures in the 'Packed Lunch' podcast; and the Schomberg Center for Research in Black Culture at the New York Public Library also podcasts its lectures. Recorded lectures are an easy means of podcasting content and indeed they make up a large majority of academic and research content (Brabazon, 2012: 143), but different universities are investing in more complex formats.

To review, a podcast in its most basic sense is a transportable, digital audio file downloaded from the internet and played on an MP3 player, smartphone app or computer. Podcasts were originally conceived out of a university research centre in 2003, where initially their creators were interested in the subversive and disruptive qualities of the medium. Some, like David Winer, believed that the medium would be dominated by amateurs, much in the same way as blogging (Walker, 2015), but it is also a medium where a full spectrum of professional broadcasters exist – from big media organisations like NPR, the BBC and CBC, to universities and research organisations like MIT and LSE.

Podcasts from academic and research organisations, though dominated by the standard lecture, are becoming increasingly varied as universities and funding bodies invest more in diverse forms of dissemination in order to react to audience trends and interests. As we show in the next part of this chapter, podcasting can fit into other parts of the research cycle too, not just in the dissemination stage. But before moving into the ways in which you can use podcasting as an academic or research professional and the different formats, we'll first summarise the advantages of the medium.

Why you should podcast your research or project:

1   Podcasting helps you reach wider audiences
2   No topic is too niche: riding out the long tail
3   Podcasting fits easily into the Research Lifecycle
4   Podcasts are research
5   Podcasts can be cheap and easy to make

5.3.1   Podcasting helps you reach wider audiences

As shown in previous chapters on blogging and social media, digital engagement can help grow your audiences beyond the confines of academic journals and those research communities already plugged into the literature. Likewise, podcasting puts your research on a completely new platform, increasing the odds that new audiences – from politicians to lay persons – will hear about your work.

Todd Landman is Professor of Political Science at the University of Nottingham as well as the creator and host of *The Rights Track*, a podcast on human rights, in which he aims to 'raise awareness about human rights analysis for students, academic researchers, policy makers and practitioners working in the field of human rights' (Landman, 2012). Landman has had such a positive and rewarding experience that he is surprised more academics don't podcast. In a 2016 article in *The Guardian* newspaper, he commented that as part of a vast network of scholars producing evidence on human rights, and whose work has been communicated through articles, books and reports, he reflects that 'I am limited in my ability to reach the people I would most like to engage and influence – those who do not have an academic understanding of human rights but might benefit from finding out about it'. Podcasting, he says, has given him a unique tool for engagement:

> For me, ... The podcast format is like a fireside chat – it allows listeners to hear experts discuss their work in their own voices, and allows the experts to express themselves more freely than in the usual academic forms of dissemination. We have even been able to work in questions from social media to provide real-time responses within our podcasts. (2016)

But just how far do they travel? Data on who, when and how audiences access and listen to podcasts are notoriously hazy. Podcast metrics don't stretch as far as a download, but there are other ways to derive a granular glimpse into your listenership. Midroll, which produces dozens of podcasts, from *WTF with Marc Maron* to the weekly podcast from *Science* magazine, polls its listeners to get a sense of audience demographics. What it found was that podcast audiences are young, highly educated and diverse. Of the audience, 67 per cent were aged 18–34 (Midroll compares this to 30.2 per cent for terrestrial radio). Additionally, 58 per cent of the surveyed audience had a bachelor's degree or higher (Riismandel, 2015).

At the LSE, we could pull up statistics for specific episodes from those who accessed our podcasts through our WordPress blogs. For the *LSE Review of Books* podcast, we produced a three-part series of podcasts from Brazil showcasing LSE's research impact there. The podcasts focused on issues around urban planning, crime, oil and social work – all based on interviews and discussions about academic papers and conferences by LSE academics and researchers. Using Google Analytics to search for both the URLs of our blog posts, which had the podcasts embedded, and the internet networks used by audiences to access this content, we were able to get a more detailed look at our audience members who came from large institutions (though unfortunately this excludes individuals who may have accessed from home).

The results were surprising. The *LSE Review of Books* Brazil podcast series was listened to widely in policymaking circles in the US, with access stemming from the Department of Energy, the trade union AFL-CIO, the New York City Department of Planning, and NASA. It had even been listened to widely in Brazil and around the world, where individuals from the

São Paolo and Rio de Janeiro state governments had accessed the podcast through the blog as well as the New South Wales Department of Education and Communities, to name a few. Not only did our reach extend farther geographically than we could have imagined, it had penetrated some key government and policy offices within those countries (see Figure 5.2).

Figure 5.2 Cheryl Brumley interviewing former mayor of Colombia Enrique Peñalosa in Brazil in 2013

Credit: LSE Cities

Nigel Warburton, formerly Senior Philosophy Lecturer at the Open University, now a freelance philosopher, was host of the hugely popular *Philosophy Bites* podcasts, an interview series that covered everything from the great thinkers like Plato to language and truth. The podcast was one of a handful of mega-hits in the academic podcasting world and spawned another series of podcasts, *Social Science Bites*, sponsored by SAGE, and a number of books written by Warburton and the show's producer, David Edmonds. Since its premiere in 2007, *Philosophy Bites* has received over 20 million total downloads. Warburton said, 'This was almost as many downloads as the whole of Oxford University's audio/visual content' (Brumley, 2014: 3).

5.3.2   No topic is too niche: Riding out the long tail

Warburton's podcast, and also highly produced radio shows turned podcasts from NPR, the BBC and podcasting companies like Midroll and Gimlet, have dominated the iTunes charts. These podcasts generally have high production values, as evidenced in aspects like their scripting or sound design, which can be a daunting listen for those thinking of starting their own project. However, they sit at the head of what writer Chris Anderson would call the 'long tail' (2008; see also Berry, 2015: 172), and on the actual tail end of digital audio offerings sit a plethora of novice and niche podcasts (see Figure 5.3). Whereas Warburton's podcasts on philosophy struck a chord with a more general audience, the world of academic podcasting is not merely a broad topic space.

Anderson's 'long tail' is applied to the retail economy but the theory is widely applicable to internet-based content. 'In the long tail marketplace, a small number of traditional hits still dominate at the head of the curve, but they now compete for attention with an increasing array of niche products, populating the tail,' write Markman and Sawyer, in their article on independent podcast makers and their motivations (2014: 25). They continue:

> Anderson argues that access to low-cost (or no-cost) production and distribution tools has created a new class of producers, frequently technophiles or innovators, who can exploit the economics of the long tail by marketing to a specialized but geographically dispersed audience. As a result, independent podcasts situated in the long tail can offer a more diverse range of audio content than traditional broadcast radio.

Although it is the big hits that dominate the headlines and garner attention for the medium in the wider press, the podcasting space is, in fact, largely comprised of amateurs providing narrowly targeted content and who sit on the 'infinitely thin end' of the 'long tail' (Berry, 2015: 172), creating content more tailored to smaller but interested or, to some, influential, audiences. Markman and Sawyer, in identifying the motivations of independent podcasters, found that those motivated by the 'long tail' factor produced podcast content because 'they liked the convenience of the medium, freedom of the medium, were interested in filling a niche/un-served market' (2014: 27).

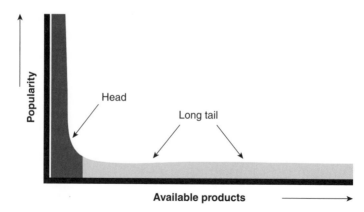

Figure 5.3   Illustration of the long tail (adapted from an image by Cheryl Brumley)

Not every one of these podcast makers intends to create the next 'breakout hit' like *Serial* (Carr, 2014), but the sheer number and variety of podcasts signal that there is an appetite for variety and, although audiences can be modest compared to the podcasts at the top end of the tail, the potential remains for a niche academic podcast to reach a rewardingly sizeable audience. This chapter provides examples of a full range of these 'niche' topics, from immunology to pedagogy, all of which have found captive audiences inside and outside their university.

5.3.3 Podcasting fits easily into the Research Lifecycle

Podcasts can be incorporated seamlessly into different stages of the Research Lifecycle, no matter what theme or topic you are researching. We can easily imagine podcasts created at the final stages of your project as part of the communications strategy or content dissemination, but academics and researchers in all fields have many opportunities to share findings and data at those early stages too. The idea of broadcasting the building blocks of your research so early on may be somewhat off-putting, especially for early-career academics sensitive to criticism, but there are many potential rewards and opportunities that spring from starting podcasting near the beginning of the Research Lifecycle.

Let's first take the example of *Researcher in the Field*, a series of video podcasts (with picture slideshows) featuring interviews with researchers and clinicians working in the area of immunology in Kenya and Malawi, created by Paul Garside, a professor of Immunology at the University of Glasgow (2013) (see Figure 5.4). The podcasts were born out of discussions in his department on how best to use his time on sabbatical in Africa in 2013. His colleague Alexandra Mackay came up with the idea of using specially created multimedia content to increase public engagement with and awareness of the project. Furthermore, the podcasts would act as a record of Garside's three-month stint in Africa, and perhaps create those first stepping stones towards impact. The result was an eight-part series with interviews covering everything from rheumatoid arthritis to health outcomes in the Mathare Valley in Nairobi.

Garside found that podcasts were a good ruse to break the ice with researchers he met in the field, and found the logistics of setting them up to be remarkably easy. Before doing the podcasts, he simply thought about what the interlinked stories might be and what messages he most wanted to get across. He then chose guests based on their role in their organisations and whether or not he thought they would effectively and enthusiastically communicate their research and roles within those organisations.

Garside said the challenges associated with interviewing people of varying backgrounds from another place could be mitigated by the way you use language:

> You have to be concise and relatively simple without appearing to be condescending – people [you interview] may not be highly educated but that doesn't mean they are not intelligent. Having an 'in' (some sort of common link) works and football worked well in Africa for lots of reasons. (Garside, 2016)

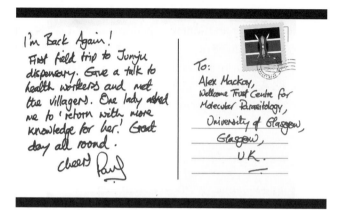

Figure 5.4   Screenshot from Paul Garside's *Researcher in the Field* video slideshow podcasts

Credit: Professor Paul Garside, University of Glasgow

The web page that hosted the podcasts was one of the most popular on the department's website, receiving 600 unique hits in one year, with an average of three minutes spent on the page. Garside said he also received positive feedback from the academics and research-ers who featured in his podcasts both in and outside the UK. He had even received several emails following the podcasts from researchers who said they came across them online and were encouraged to reach out to him to collaborate on projects. (You can listen and watch here: www.gla.ac.uk/researchinstitutes/iii/wtcmp/field.)

5.3.4   Podcasts are research

But what about podcasts as pieces of research in themselves, separate from the data collection and dissemination parts of the Research Lifecycle? While producing the Brazil series of podcasts for the *LSE Review of Books Podcast*, Cheryl Brumley, co-author of this book, interviewed renowned criminologist Silvia Ramos about violent crime in Rio de Janeiro. The interview would feature in a 30-minute podcast on how Brazilian NGOs were transforming the lives of young people in violent communities. (You can listen here: http://blogs.lse.ac.uk/lsereviewofbooks/2014/04/03/lse-review-of-books-in-brazil-episode-2.)

During the interview Ramos commented that Brumley was utilising an 'investiga-tive methodology' and, by doing so, she was producing a piece of research that could stand on its own. Not only had Brumley talked to Ramos for this particular podcast on crime, she also spoke to local NGO leaders, academics and even circus performers. Cheryl explains:

I never really thought about podcasting as creating a distinct piece of research. With my background in radio reporting and producing, I was simply asking questions, and piecing together parts of a story in the way that I always did. In this case, I was looking into how deprived areas of Rio benefitted from community arts programmes. I had never previously thought of it as an academic activity. But Ramos was right. By asking questions, and delving deep into the facts on the problem of crime in Rio de Janeiro, I was stitching together a story that was in its own way, a piece of research: how to stop a cycle of violence in the urban space. The sciences and social sciences at their core are about putting together the pieces of a puzzle. This experience therefore got me thinking about podcasting as more than just a means to an ends. The microphone forms a sort of contract; you listen, collect and clarify. The interviewee in turn understands they must deliver opinions, facts and, in some cases, emotional responses to your questions.

Why can podcasts help you to achieve that? The perception is that a microphone makes people tense up, and for the novice interviewer and interviewee this is sometimes true. But with many interviews, the longer they go on, the more likely that the initial awkwardness and nervousness will assuage. A microphone and recorder in this case transform into forgettable accessories. As the interviewer and interviewee settle in, mutual learning can take place. And you can use the content for more than just qualitative ends – you can broadcast it. This is a really beneficial tool for academics to 'do research'.

Utilising investigative methodologies forces us to ask more hard-hitting questions in a way surveys could never do, including questions that are more open-ended. Additionally, an investigative methodology is particularly suited to research on difficult topics, especially in the area of psychology. A team of researchers at the Centre on Behavioural Health at The University of Hong Kong write:

> Although it [investigative methodology] does not preclude other techniques for gathering data, investigative research tends to rely on disciplined, naturalistic, and in depth observations over a time span in diverse contexts. It is particularly suitable for uncovering, understanding, and reporting social phenomena that may be hidden from or not easily accessible to observers. (Yau Fai Ho et al., 2006: 17)

In essence, researchers, acting like journalists, may uncover truths unapparent in surveys and other more traditional quantitative and qualitative research methods:

> Investigative researchers have good reasons to delve into open secrets, insiders' views, rumor, remarks made off the cuff, and activities behind the scenes; to treat official pronouncements with skepticism, and to tap informal networks of communication. ... Diverse sources of data may be used, including observations, mostly unobtrusive, of everyday experiences; and unsolicited or spontaneous utterances by, and conversations with, insiders as well as outsiders of the target group; records, documents, and archives in public or private domains. (Yau Fai Ho et al., 2006: 25, 34)

The podcast interview presents the perfect excuse to probe, question and react to the responses of research participants with the goal of including the more personable, face-to-face stories that have traditionally been left out of research methods. Obviously, using podcasts as an extension of an investigative methodology is more effective and practical for some areas of studies than others, namely, psychology and sociology, but it nevertheless leaves open the possibility for other disciplines to include more human-centred methods of research.

5.3.5 Podcasting can be cheap and easy to make

Later in the chapter we will discuss the many different types of podcast format and the labour involved in making each. The ones stripped back to their most basic – the monologue format, the interview and the lecture podcast – are the least intensive and require a minimum amount of entry skills. Most importantly, podcasting is a forgiving format – you learn as you go along. Yet it is still worth investing time in the fundamentals of the medium.

We still strongly encourage all potential podcasters to familiarise yourself with the basics of recording and editing before setting out on your own project, but these technical aspects shouldn't be seen as a big barrier to taking the crucial first steps.

*****

Despite all the potential advantages already discussed, podcasting is often overlooked by researchers and academics. Many in universities are put off because there is often not enough time between managing tight academic publishing schedules, attending conferences and essay grading to learn a new skill, let alone one that requires a touch of technical know-how. For communications professionals and team leaders in NGOs and charities, tight budgets and limited enthusiasm for trying new forms of media are often barriers. Technical equipment comes with its own alienating lexicon that can be discouraging and intimidating to an already overworked person or team.

But here is the good news for the scared or the sceptical: podcasting is inexpensive and, depending on your audience's expectations, it can be a surprisingly non-technical enterprise. For those who have the time and budget to experiment with the more complex styles of narrative podcasting and sound design, it can be a time-consuming but rewarding experience to bring in many voices and guests. At the other end of the spectrum, simple debate-based or descriptive podcasts with just one or two voices will be suitable for the majority of projects that just seek to make public aspects of one's research with as little hassle as possible. With a recorder, a microphone and, often, free editing software like Audacity or Garageband, you have everything you need to begin to podcast. More dynamic software like Avid Pro Tools, Adobe Audition, Hindenburg and Reaper range in price from US $60 to $400 (that's roughly £40–£270), but are only necessary for the podcaster requiring the amenities for a more complex format. Before we get into what those formats are, it is important to remember to start small. You can hone your skills as you progress forward.

We hope you remain much more convinced of the many merits of podcasting for sharing research. Researchers looking to reach wider audiences outside those already in tune with the usual ivory-tower publishing channels can do so through podcasting. Podcasting is also useful as a tool throughout the Research Lifecycle. The audio-based medium presents a means to distinguish yourself from the crowded and competitive research landscape. At the same time, it is an accessible and easy means of achieving all these ends. After looking at the reasons for podcasting and visiting case studies from a range of researchers, section 5.4.2 will go into more detail on the types of podcast format and the technologies and skills that are involved with each.

5.4 HOW TO CREATE A SUCCESSFUL PODCAST SERIES

5.4.1 The questions to ask before you record anything

Before recording any audio and sharing anything for your project, taking the time to ask yourself and your team some key questions about your audience and what you want your podcasts to achieve is essential. Not only will it give you the space to think critically, but you will have a plan to return to at future stages and with which to make sure your podcasts are doing the job you originally envisioned.

- What do you want to achieve?
- What subjects and content do you have access to?
- Who are your target audiences?
- Is podcasting your research really necessary?
- What ethical considerations do you need to keep in mind?

Question 1: What do you want to achieve?

Why do you want to start a podcast? Is it a means for your organisation to expand its digital presence? Are you a fan of podcasts and want to try it for yourself? Do you want to reach non-academic audiences? Being clear from the outset about why you want to create a podcast series will help focus your project.

Co-creator of the podcast *Pitch*, a self-styled 'narrative podcast about music', Alex Kapelman encourages would-be podcasters to make a mission statement, noting also that the mission statement may change over time. He says:

When I started 'Pitch' with Whitney Jones, my mission statement could have read something like: I'm making a podcast because I want to improve my skills, gain experience, and
eventually get a job at a show like *This American Life*. As of this publishing, it's probably something more like: I'm making a podcast because I love reporting the types of stories we do on
'Pitch,' and I want to find a way to have an income from being a radio producer. (AIR, 2016)

Question 2: What subjects and content do you have access to?

Potential podcasters hoping to make use of interviews need to consider the access they
have to their ideal guests. Maybe you are a PhD researcher with a foot in the door to some
influential academics within and outside your institution? Or perhaps you are a knowledge
worker who regularly travels and you want to broadcast the conversations you have with
colleagues abroad? If you want to start a podcast, try to think of how it can be incorporated
into your existing working life around the people you encounter every day. Podcast what
you know with the people you know first. From here, your network will expand to include
those on the periphery of your contacts list. Future guests will be all the more encouraged
to take part by seeing a long repertoire of people who took part before them.

Question 3: Who are your audiences?

Who do you want to listen to your podcasts? The answer to this question will inform
your approach and will help determine the format and time required to produce them.
Different audiences have different expectations, different time schedules (and they
also use different consumption platforms, which will be detailed later in this chapter).
For example, niche audiences, or 'long tail' consumers, are generally tolerant of long,
unedited, lecture-style academic soliloquies. If the general public is your target audience,
keep in mind their attentions are harder won.

Question 4: Is podcasting your research really necessary?

After asking yourself the previous three questions, this might seem redundant. However,
maybe to achieve your aims, make use of your access and reach your intended audiences,
you don't need to produce a podcast at all. Will a series of blog posts, or a video, or a data visualisation meet your audience's expectations and tell your research story more successfully?
Will it work on social media? The answer to any of these questions may very well be yes, or
you could use a combination of the other formats covered in this book. Creating a podcast
series just for the sake of it can be counterproductive to your ultimate aims and it can reduce
the impact of your research. Furthermore, it may make for a tedious podcast for the listener
if you are just podcasting for the sake of it. Not everything is meant to be heard rather than
read. Remember that a poorly executed podcast can be worse than producing no podcast at
all. Don't let this realism put you off the medium itself; rather, it should help you determine
the digital medium that best helps you achieve your digital engagement objectives.

If you are to be honest with yourself about your own interest in audio as a medium, it follows that you should also be realistic about your time. You may want to be the next academic podcasting star, with a serious interest in upskilling, but you can only do what your schedule allows.

Question 5: What ethical considerations do you need to keep in mind?

Most researchers will be familiar with ethical considerations. Ethics committees are an enshrined practice in the modern-day research environment. Those ethical obligations carry over into the podcasting sphere too and they are all the more pressing for the person who aims to podcast from the field, where research participants are being used as podcast interviewees. In the same way researchers should consider the impact that any dissemination practice has on their participants, so should the academic podcaster. The British-based Economic and Social Research Council (ESRC) (2016) advises researchers when considering any public engagement initiative to consider its

> possible impact on research participants, their families and associates, organisations, and populations from which the sample is drawn needs to be thought through – particularly where anonymity may be jeopardised or where there is potential for stigmatisation of individuals or groups, or misuse or misrepresentations of research findings (e.g. to further political agendas).

Your participants should be made aware of where you intend to publish your podcast and how you intend to use it. Even if you have their explicit consent, your own discretion is important. If you detect there could be any negative outcome for them as a result of their taking part, you should consider disguising their voice, or not using them at all.

5.4.2   Different types of podcast

In this section, we will look at the different types of podcast format and the implications each have on the skill set you need to acquire and the time needed to produce individual episodes. Not all podcast formats are created equal in terms of the demands they place on busy academic schedules and this section is meant to provide the information you need in order to choose the podcast that best suits your ends.

Different formats require a different level of skill and also varying levels of time commitment. It is hard to give an exact estimate of how long a specific podcast format would take you to produce because of the number of variables involved (for example, the length of your recording, the speed at which you edit, whether or not it is mixed to music). In Table 5.1, we outline the types of format available to you as a research professional, examples of podcast series which utilise each and the factors which you need to consider before embarking on your project.

Table 5.1 Different podcast formats

Type of podcast	Best for	Factors to consider
Monologues – this involves a single voice either reading from a script or delivering an extemporaneous speech. Podcasts like these resemble Winer's early podcasts, though they are less common now Example: Memory Palace, Viva Voce podcasts	– The creative scriptwriter – Audio diarist	– Monologues that require a script will ideally be read two or more times to give you more options when editing – Audio diaries benefit from ambient sounds from in the field to add an interesting and added dimension of sound
Interviews – this involves a discussion between two or more persons in a volleying, question/answer format (Examples: Philosophy Bites, Teach Better)	– Exploring single issues – Exploring topics on which you are not the expert but want to find out more – Growing your network of scholars and researchers	– Editing any podcast of length down to 10 or 20 minutes is a big task if you talk for an hour and a smaller task if you talk for those exact lengths – Who do you have access to? – What questions will elicit the most exciting responses from your guest?
Lectures – this is a straight, uninterrupted recording of a lecture or presentation. It is the easiest to produce and the most recognisable of formats	– For the time-strapped academic – Keen lecturer	– You don't have to include every minute of your presentation or lecture. If, in retrospect, you thought a part of it was too long, or sagged, edit it out – If your presentation was highly reliant on a slide pack, purely using the audio may be confusing or hard to follow
Radio-style magazine shows – this type of podcast is characterised by having two or more distinct segments and can involve a combination of any of the above formats (Cited podcast, *LSE Review of Books*, Audible Impact)	– Researchers with institutional backing – Researchers with a serious interest in the medium of audio or those with an already established technical background	– The time required for these podcasts is more intensive and will require more resources, including time and possibly staff – Musical interludes between segments gives the audience a break and some time to absorb what's already been covered
Scripted narration – scripted, narrated shows resemble radio reports. You, as the producer, would conduct an interview either in person or over the phone, and then afterwards, use an editing program to break the audio into clips. With these clips you would then script in and out of each	– Researchers with institutional backing	– Scripted narration allows for greater control – But it will require more of your time. Depending on the length of your raw audio from your interviews and the desired length of your podcast, developing a script and executing the finished product could take a few hours to a few days of work

Monologues

Monologue formats are not as common and they require a host with a certain degree of confidence and commitment. The risk with the monologue format is that the ear can grow easily tired of the same, uninterrupted voice. Music, ambient sound and clear scripting can all help these kinds of podcasts grab the attention of listeners and attempt to garner a loyal following. There are a few examples inside and outside the academic world of podcasts that utilise this format.

The Memory Palace podcast, hosted by Nate DiMeo, joined the Radiotopia podcast group in 2015 and is a great example of a creative use of monologue. Each episode is narrated by DiMeo and mixed to music and archival sound. The effect is a dream-like, haunting, often quirky, portrait of a person or time in history. In the transcript for the 85th episode, 'Finishing Hold', which is about a doctor accused of murdering his wife in the 1950s, we can see how DiMeo uses a direct form of narration to engage listeners:

> Sam Sheppard met Marilyn Reese at Cleveland Heights High. ... They were married in 1945: all post-war promise. They had a son. They called him Chip. And they settled down back home in Ohio, in a house by the water. ... Tack it to your wall and see it there: that perfect house and the lake, perfect family ... because it's the last clear picture we're going to get. (DiMeo, 2016)

Here DiMeo is economically raising your investment in the characters before introducing a tragic twist:

> They were entertaining that night. Their neighbours came over for dinner and drinks and left at some reasonable hour. Then the Sheppards watched a movie on TV for a while until Sam nodded off. Then Marilyn put Chip to bed and then herself. And either Sam went upstairs and brutally murdered Marilyn while their son slept or he awoke to cries of help, rushed to the bedroom and found a shadowy figure assaulting his wife. (DiMeo, 2016)

Monologues without the aid of narrative arc or script can easily bog down a listener. However, DiMeo's podcast is captivating partly because it addresses the listener directly, drawing them into a story. He doesn't attempt to take you on a journey through the entirety of someone's personal history. He aims for a more general audience, one that, in the case of the Sam Sheppard podcast, is not required to have a specific interest in historical murder mysteries, merely an interest in a good story.

Historical subjects, like others in the social sciences, come with their own multitude of facts and figures. Instead of using podcasting to convey everything he knows about a subject, DiMeo strips each topic down to the essentials. The *Memory Palace* exists not only to inform; it also intends to evoke emotion and to entertain. The language is simple, clear and poetic. DiMeo's voice is as much a part of the character of the podcast as the subject themselves. Indeed, for a scripted monologue to work well, the host must feel that he or she is capable of reading in an engaging and natural way.

Of course, this style of writing and podcasting may not be realistic for everyone. It may not even reach the objectives that you, as an academic or a researcher, want to reach. For one, you may aim to reach a more specific audience, and you may want to foreground the multitudes of facts and figures surrounding your subject, rather than prioritising stylish language and sound design. In this case, you can do all these things through monologue, but you must be clear as to why you think using a single voice is the means for achieving your objectives. Getting the attention of listeners when there isn't an oscillation of voices between two or more subjects can be hard. This podcast format is best for those who are willing to whittle down their topic into a clear beginning, middle and end.

 Monologue formats can also be unscripted. The *Viva Voce* podcasts launched by researcher Gemma Sou rely on crowdsourced audio from researchers from around the world. On the Viva Voce website, participants can upload a 4–5 minute podcast on the topic of their dissertation or research and each podcast is accompanied by personal information about the researcher and links to other publications. Viva Voce showcases the work of academics and, at the same time, it uses audio platforms like Soundcloud and its own website to disseminate research outside the usual academic channels. Sou said she launched Viva Voce to enliven the means by which researchers share their findings with the public:

> Social science research can become sanitised when researchers are left to summarise their findings in a few lines. By literally giving researchers a voice, findings become more exciting as people are allowed to animate their findings and bring character to their research, which I think does more justice to the research that they carry out. (Sou, 2014)

Crowdsourcing audio also has the obvious benefit of growing your pool of content without placing unnecessary demands on your time.

Interviews

The interview is probably the most recognisable format and varies according to how many interviewers and interviewees you choose to have. The easiest to produce, of course, is a podcast that features a single interviewer and a single interviewee. The benefits of an interview format for the researcher/producer are many: first, you can explore topics unfamiliar to you and gain as much from the process as your listener; second, as you grow your content and feature new guests, you also expand your network of researchers and academics in your field or related fields; third, it can be one of the least time-consuming formats to produce.

The following example of two academics at Yale, who started a podcast on teaching, highlights the myriad of ways in which the interview format can enhance your research and dissemination objectives.

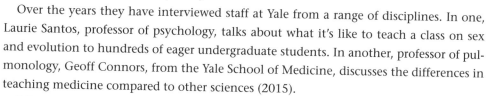

Douglas McKee, a lecturer in economics at Yale, and Edward O'Neill, a senior instruc- tional designer also at Yale, produce and co-host a podcast called *Teach Better* about teaching methods in Higher Education. Just before the two met in late 2014, Douglas had started blogging about teaching and at the same time listening to a lot of pod- casts. He thought to himself that he would like to start his own teaching and Higher Education podcast, first, because he enjoyed the experience of blogging so much he wanted to carry it over into another space ('Blogging was my gateway drug to pod- casting', he jokes) and, second, because there was a real dearth of Higher Education podcasts at the time. He met Edward in 2014, and after a series of informal chats over lunch, where the conversation would flow easily and the hour would go by quickly, Douglas thought, 'This is exactly what I want from a podcast'. He expressed this to Edward, who became emphatic: they would start a podcast together. Douglas would handle sourcing the guests and scripting and Edward would handle the technical side of things (McKee and O'Neill, 2015).

Over the years they have interviewed staff at Yale from a range of disciplines. In one, Laurie Santos, professor of psychology, talks about what it's like to teach a class on sex and evolution to hundreds of eager undergraduate students. In another, professor of pul- monology, Geoff Connors, from the Yale School of Medicine, discusses the differences in teaching medicine compared to other sciences (2015).

Where it had started by identifying a niche they could fill in the podcast market (I think Gandhi said, 'Produce the podcast you want to download,' says Edward), it became about so much more over the years. Their interview format performs a number of functions for both him and Douglas: 'It is partly about having a discussion on teach- ing; it's partly about being part of the culture at Yale; it's partly about joining in the national and international discussion of teaching and learning; and it's partly about gathering information that you're going to use later.' He also says their *Teach Better* pod- cast is research: 'I do see it as research – learning about your institution.'

Edward also talks about the importance of conversation and why the interview is an active process of knowledge creation: 'I love dialogue … Something very meaningful can happen to you by talking to someone. You learn something. You make a new piece of knowledge together. I just think that's wonderful.'

Douglas seconds Edward's thought and finds parallels in teaching: 'I find that there's teaching and standing in front of the class but that one-on-one or one-on-two time that happens during office hours, you can completely customise what you say to that student. A conversation is a really efficient way to get where you want to be.'

They have received positive feedback both from their interviewees and from listeners outside Yale. One of the main beneficiaries, they argue, is themselves: 'I think a podcast is one of my most significant accomplishments,' says Edward, 'I've learned so much about teaching from doing this show.'

Lectures

Recording and publishing a lecture is the most straightforward way of producing content. Universities around the globe produce podcasts based on their public lectures, many of which evolved out of the iTunes U platform, so successful examples abound. Individual researchers can take advantage of this format, especially if they regularly teach or present at conferences.

For those who do choose to go down the route of the lecture podcast, it's important to note the wonders of editing. Too many of these types of podcast include administrative introductions to the audience by chairpersons or even awkward silences (think about the lone cough somewhere in the rafters). Top and tail (i.e. provide a short intro and outro) these podcasts with either music or a verbal introduction, and edit out any parts of your lecture or presentation that you feel, in retrospect, failed to excite your audience.

Scripted narration

Scripted, narrated shows resemble radio reports. You, as the producer, would conduct an interview either in person or over the phone, and then afterwards, use an editing program to break the audio into clips. With these clips you would then script in and out of each (sometimes referred to as 'links').

Scripted narration allows for greater control. In the case of interview podcasts, you have to edit a single, continuous piece of audio, whereas, if you narrate around desired clips, you have more freedom to manoeuvre around content and really shape a piece. As a result, it's much easier to feature the best and most interesting parts of an interview.

However virtuous this freedom is, it will require more of your time. Depending on the length of your raw audio from your interviews and the desired length of your podcast, developing a script and executing the finished product could take a few hours to a few days of work. These podcasts, therefore, are better executed by researchers with institutional backing.

Claire Navarro is the producer of *Hold That Thought* (2013), a narrative podcast from Washington University in St Louis, the College of Arts and Sciences. *Hold That Thought* covers the breadth of the humanities and sciences. Topics range from a professor of philosophy discussing 'Claude Monet and the Science of Style' to 'The Physics of Baseball'.

Starting out, Claire was a novice to podcasts but she felt confident she could quickly pick up the necessary skills to produce them. She felt it was the 'most nimble and easier to produce than other formats and she was confident in her ability to pick up the technology'.

Claire uses a narrative style to steer and shape interviews with professors about their research. Her university has hired her to effectively use podcasting to share their research with the wider world beyond academia: 'We really try to show the depth and breadth of what goes on here [at Washington University].'

Impact so far is anecdotal but very positive. She says the professors are really happy with the podcasts. They will often share them with their own network and colleagues. Her favourite feedback was when a professor of philosophy said that after listening to the *Hold That Thought* episode, his mother finally understood his research. 'That was my favourite feedback,' she said. In fact, the podcast seems to be a hit with parents. Another parent, whose son was going to be a first-year student at Washington University, wrote to Claire to say that after listening to the podcast and hearing about the amazing research that went on there, he couldn't wait for his son to be attending the university. 'I didn't think of prospective students' parents as being part of the audience,' said Navarro.

Navarro produces three quarters of the episodes. The other quarter is produced by another member of staff. This involves picking the topics, conducting the interviews, hosting the episodes and editing them. The Dean overseeing her department, concerned that academic research is too insular, gives her much needed support. Within the university, Claire's podcasts are unique. 'Nobody else in the school is doing this regular publication that is digital. It is valuable that these researchers have this outlet,' says Claire.

Claire learns a lot about podcasting through the process of production. She has found that editing podcasts down to their essentials is the most time-consuming part of her job. 'You sort of have to be honest with yourself ... these people have been studying and working for decades and it's not possible to pick out the interesting story line,' she says. Over time, her interview process has become more focused: 'My interviews have gotten shorter. I try to not cover their whole life story.'

Radio-style magazine shows

A magazine-style podcast is one that features at least two 'segments', i.e. parts of an audio programme that are distinctive in theme. The segments can be in any of the aforementioned formats. Because of the combination of segments, which means you either have additional guests to book, or monologues or narration to script, magazine shows can require a fair amount of time to plan, produce and edit. Therefore, like single-scripted narrated programmes, radio-style magazine shows are best suited to researchers with institutional backing.

Gordon Katic is host and producer of a magazine show produced at the University of British Columbia (UBC). Previously called 'The Terry Project', which was a wider initiative at UBC for undergraduate students to learn about global issues through interdisciplinary panels and conferences, the podcast relaunched as *Cited* in early 2015 with the aim of broadening beyond the confines of UBC. The podcast is driven by Katic's desire to open up the academy, something he feels is the duty of everyone working in academia: 'There really is an onus on us to make things clear to people – not to speak in the most impenetrable jargon as possible,' he says, adding, 'The philosophy of the programme is that we aren't targeting other academics. The purpose is to tell stories about esoteric research in ways that directly apply to ordinary people's lives and to tell them in a way other people understand.'

Cited received a grant in 2016 from the Social Science and Humanities Research Council of Canada to form collaborations with different research organisations, something they had done previously, namely with a podcast on evidence-based drug policies. The new podcast season will focus on three different research clusters: criminal justice reform, income inequality and social justice, and climate change.

Cited has a very clear vision of how it hopes to mobilise knowledge. 'The philosophy of the programme is that we aren't targeting other academics,' insists Katic, 'The purpose is to tell stories about esoteric research in ways that directly apply to ordinary people's lives and to tell it in a way other people understand.' One way he and other members of the *Cited* team can achieve this is through co-production. The podcasts won't simply reveal research findings to different publics; they will ensure these publics have a voice in the podcast. 'We're trying to tell stories about researchers embedded in the community,' says Katic. 'We respect other types of knowledge that will expand the epistemological boundaries of academics'. In practice, that means the voices of activists and other previously excluded 'lay persons' will sit alongside experts.

Cited is also a functioning research project. As part of the grant, Katic and his team will be studying themselves. Part of this will be to form an advisory board that will evaluate *Cited*'s different seasons. The other part is that it will have a researcher documenting and evaluating how effectively it mobilises knowledge. 'We needed to evaluate ourselves in real time,' says Katic, and part of the reason is so that they can pivot in case their strategies aren't working. Taking inspiration from the Gimlet Media podcast *StartUp*, where, in the first season, the host Alex Blumberg documents his journey as CEO of a new podcast company, of which the first podcast was *StartUp*, the *Cited* team will be documenting their journey of mobilising knowledge through podcasts in their own podcast. 'That's what is good about podcasting,' says Katic, 'people are transparent and it's personal and they are honest about what they are doing.'

Katic notes that not every academic or knowledge worker is equipped with the will or the means to create really powerful stories through audio. He says often people are surprised that he works in a team with upward of five members. But he insists that if you want to create something closer to modern-day podcasts, that is, more aurally complex, story-focused, radio-style broadcasts, a professional, full-time staff is a necessity. 'Academics are starting to realise the power of storytelling,' says Katic. 'The problem is very few people have the infrastructure around them to support them in doing that. Journalists already know how to tell good stories; why should we put the onus on journalists to have someone support the storytelling aspect?'

5.4.3 Hiring a professional to create your podcasts

Once you have an idea of what podcast format you'd like to feature or how you'd like to incorporate podcasting into your Research Lifecycle, the next thing you need to decide on is whether you want to hire a professional to create this content for you, or whether you'd like to go down the DIY route. Now is the time to be realistic about your schedule, time, and also your interest in the audio medium.

We would recommend using professionals to create your podcasts if you are looking for a highly polished finish, if the story you want to tell is complex, and especially if the format is completely new to you and you have little to no experience of listening to podcasts or radio shows. The podcast benefits from the best practice traditions of radio, and, therefore, although it is a new medium, it has undergone over a century of professionalisation. In the same way we asked you to consider hiring a professional to carry out sophisticated data visualisation objectives, likewise, if you have lofty goals for audio, you may want to consider hiring a producer/editor or producer/journalist. We will note the difference between the two in a moment.

If podcasts are completely new to you or you feel that editing audio files is something you'd like to hand over to a professional, it could be best to hire the skills of an experienced audio producer/journalist, or at the very least talk to someone you know who works in radio or audio media who can show you examples, provoke questions around what goals you'd like to achieve and help you flesh out exactly what you want to create.

What to look for in a podcasting partner

Depending on your desired aim, you will need to hire either a producer/editor or a producer/journalist. 'Producer' is a catch-all term for someone involved in an aspect of podcast or radio production. Since radio budgets are generally more modest than their TV counterparts, this term encompasses a huge range of skills, from recording audio, to directing 'reads' for scripts, to editing. However, some producers specialise in certain areas. Producer/editors have a sound editing focus whereas 'producer/journalists' have a reporting background and are skilled at shaping a narrative and conducting interviews. So if you want to hire someone to produce a lecture podcast, you will need a producer/editor, and for a narrative podcast or magazine-style show, ideally you will hire a producer/journalist. The distinction may not be so clear from the outset but it will be once you hear their work.

You should ask potential hires for a 'showreel' – a collection of work they have produced, edited or reported on. When listening to these showreels, use your discretion: does the audio have the quality objectives you desire? Have they produced podcasts in the format you desire? Do they have a specialty in the social sciences or hard sciences that might be of benefit to you?

When you finally locate the right partner, you will need to give them clear ideas and examples of audio styles that you like or particularly don't like. For example, bringing along examples of podcasts or radio shows produced by competitors or colleagues at other organisations and describing any aspects of the editing or sound design you'd like to recreate can help the team to get a much clearer idea and produce something closer to what you have in mind. Storyboards or script ideas can be valuable at this point too.

In most cases, having a budget for the creation of these types of media will be the deciding factor on which route you take. Podcast producers often deal with clients who are ignorant of just how labour-intensive highly-produced projects really are. It is important, then, that you understand how your freelancer intends to utilise their time, and to allow them to outline for you the resources they need in order to deliver you a finished product that meets your expectations. If they are asking for too much money or time, it is most likely an indication that you, as the commissioner, need to temper your ambitions.

5.4.4　Tools for making your own podcasts

If, after considering your other options, you decide to go the DIY route, you're in luck: podcasting provides a range of options for different skill sets which means you can still achieve a high-quality production as a time-poor and budget-stretched researcher.

We can imagine that podcast formats range from the simple to complex on a spectrum, starting with lecture recordings and monologues on the more simple and easy-to-create side, ranging through to journalist-style packages and the scripted magazine-style format on the more complex side. It follows that the types of equipment used to make these podcasts also fall on a spectrum. Relative to other multimedia, audio is cheap and far more approachable for the technically uninitiated. But, however easy, it is worth putting in the time to learn how to collect good quality sound. 'Don't make the mistake that so many academic podcasters make of thinking that it's ok to put out unedited audio,' said *Philosophy Bites* host, Nigel Warburton, in an interview with LSE. 'It's the most tedious thing to listen to and undermines your whole status as creating something worth listening to. ... Think [instead] like a radio editor and producer' (Brumley, 2013).

Given the increasing presence of well-produced podcasts and slick radio programmes, ensuring that you collect the best possible sound will help your audio project stand out. Free versions of audio-editing software, like Garageband and Audacity, will suit *all* the needs required by the casual editor and *most* of the needs for those editors-on-a-budget who are seriously into their sound.

Choosing a microphone

The right microphone is the key to good sound. Microphones can come in the form of attachments to phones, in-built mics, lavaliers (clip mics), stationary podcast mics with tripods, or a hand-held one (see Figure 5.5). The quality of sound picked up varies widely for each. For speech recordings, it is best to use more targeted recording devices, meaning those with narrower polar patterns that pick up the voice directed at it rather than all surrounding noise. Stationary devices, which sit between the interviewer and interviewee, may allow for hands-free recording but unless you have a completely sound-proof space these devices will pick up a surprising amount of noise, from chair squeaks to air conditioning. Most of these noises can be forgiven, but too many distracting clangs and bashes will turn listeners off. Likewise, mobile recording devices are small and portable, but as the quality is poor, they do not come recommended for lengthy interviews.

Keeping the microphone in your hand, or on a stand directly in front of the person speaking, will allow for more targeted voice collection. Voices project differently and have varying levels of robustness. Before launching into an interview, test out the right microphone distance by getting your interviewee to talk a little while you determine the right distance for their voice (industry standard is asking what he or she had for breakfast, which will give you a nice collection of that portion of your interviewees' daily routines!). It is best if the mic is held at a 45-degree angle to avoid popping noises on p and b words. If you hold a microphone too close to an interviewee, it could distort the audio; if it is held too far away from them, you risk picking up unnecessary ambient noise. The right balance is easily struck with some practice.

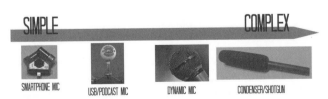

Figure 5.5 The microphone spectrum

Credit: Photo of USB Mic from Tod Baker (CC-BY-2.0), Flickr.com; Photo of Shure mic from Arbyread (CC-BY-2.0), Flickr.com

Additionally, flat surfaces reflect sound, which can result in hollow and echoing speech. If you have no choice and you have to record in a large lecture hall next to hard surfaces, throw your jacket, scarf or backpack/rucksack on the table, as this will help to absorb some of the sound.

Choosing a digital recorder

Your tablet or laptop will probably have an in-built microphone that can record straight on to editing programs or you can record directly into your program using a stationary podcast mic that can sit atop a desk or workspace. The use of a recorder, however, comes in handy for interviews in the field. Like microphones, recorders exist on a spectrum of the simple to the very complex (Figure 5.6). Differences in sound quality are not as easily detectable, and one can get away with using a low-cost digital voice recorder such as a standard Olympus model, starting at $80 or £50. Some recorders work better with a particular type of microphone. The Transom.org website, a non-profit focusing on radio production training, has in-depth information on these ideal microphone–recorder pairings, especially for researchers looking to take their equipment mobile (Towne, 2012).

As podcast equipment becomes increasingly affordable, easy graduation from novice equipment to more serious forms of recording becomes all the more possible. If you are simply dipping a toe in the water, start by purchasing very basic equipment and move up to the more complex as you find yourself more committed to the medium.

Figure 5.6   The digital recorder spectrum

Credit: Photo of Laptop from Max (CC BY 2.0), Flickr.com; Photo of Sanyo USB recorder from peter-rabbit (CC BY 2.0), Flickr.com; Photo of Tascam recorder from Scott Schiller (CC BY 2.0), Flickr.com; Photo of Zoom H4n recorder from Leon Cych (CC BY 2.0), Flickr.com

For further help in choosing the right equipment, consider these four points:

- **What is your desired quality level?** Would it benefit to move towards the higher-quality end of the equipment spectrum or are you just a casual podcaster?
- **What recording environments are available to you?** What is the ambient noise like? This will help you determine how narrow you want your microphone polar pattern to be (cardiod? omni-directional?)

- **How long do you want your podcast to be?** If they don't extend over five minutes, quality could be less of a concern than if extended interviews are featured.
- **What types of interaction will your podcasts consist of?** For example, your voice only, an interview, or a conversation between a number of academics? If more than two people will feature, or if you are simply delivering a lecture, a stationary podcast microphone/recorder is the most ideal device (despite the amount of ambient noise it picks up).
- **Does your university offer any organisational support?** Some universities will lend you equipment or the use of a studio. Find out from your communications team or through learning technologists on your campus.

Editing

The right microphone paired with the right recording equipment will greatly improve the quality of sound you collect. In turn, this improved quality will make for a smoother editing process. Many first-time podcasters avoid delving too far into the seemingly complex landscape of sound editing. At its most basic, editing involves cutting and deleting. The other, more complex, processes involved shouldn't discourage the absolute beginner. Even a little time spent cropping your podcast down to the essentials and cleaning up any distracting flaws will go far in making your podcast sound like a professional production. Editing software is generally built for music producers, but new software, like Hindenburg, is tailored more to the podcast and radio producer and will allow you, for instance, to record Skype conversations directly into the program and to organise clips in a board to the side of your editing workspace.

Audacity is a free and widely used programme that works on both PCs and Macs. The one disadvantage of *Audacity* is that it is 'destructive' editing software, meaning once you make a change, like cutting a piece of audio, it is permanent (unless you continuously undo your edits until you're back to the original cut). Other software lets you reverse these changes at any point.

5.5 SHARING YOUR PODCASTS ON AUDIO PLATFORMS AND SOCIAL MEDIA

Once you have your podcast edited, it is now time to upload it onto different audio platforms. Here we look at the different platforms available to you, before noting some best practices for sharing your content. In terms of getting a wider audience, podcasts face different struggles from those of blogs and more visual media and, as such, sharing podcasts online requires a social strategy responsive to these differences.

5.5.1 Sharing podcasts on audio platforms

To ensure your finished product reaches the widest audience possible, consider rolling out your podcasts on several different audio platforms at once. For instance, Claire Navarro, producer of *Hold That Thought*, uploads the podcast on iTunes, Stitcher, a 'podcatching' app, and PRX, an online marketplace for audio content.

Soundcloud has become increasingly popular and allows for greater interaction with audio than iTunes via comments and shares. It also provides you with embed codes so you can upload a player on your blog, website or social media platform. The players give embed options for flash-supported browsers or HTML5, but note there are some reported issues with these players appearing on iPads and iPhones. If you're uploading more than a few hours of audio, you have to pay a subscription fee.

Additionally, you might want to check if your university or organisation will upload your podcasts to their iTunes account and other platforms. Most universities have well-established followers and RSS subscribers and if you are lacking in online visibility, this is most likely the best way for you to develop regular listeners. At LSE, we were able to benefit from the already existing followership both on Soundcloud and iTunes by branding all our podcasts as LSE podcasts, and distributing them through LSE's podcast channels. The advantage is two-fold. First, for most university podcasts, the greatest challenge is gaining the initial podcast numbers to sustain and grow your audience over a long period time. But if you're able to get on your university or organisation podcast channel, you're already a few steps ahead. Second, you're associated in name with the international name of your university or organisation, so you're a more trusted source. However, the disadvantages are also clear: you may have to 'tow the company line' if your message does not align with the brand or identity of the organisation, and thus sacrifice your independence. However, whether you are publishing through your university or independently, you are responsible for what you say regardless of where you publish. Academics and researchers should exercise the same caution in uttering a possibly slanderous sentence as they would a libellous one.

5.5.2 Sharing podcasts through social media channels

Why doesn't audio go viral? It is a question that has dogged the podcasting community for years. Podcasters look in envy as other media, such as blogs, photos and data visualisations, spark a 'Twitter storm' or result in hundreds of shares on Facebook. But audio has a tough time engaging the digital world. Even *Serial* couldn't solve this problem. 'Audio remains resistant to being as rapidly transmittable across the Internet as static or video visual memeography,' says industry expert Nick Quah (2015b). Nate DiMeo of *The Memory Palace* also writes: 'Audio never goes viral. If you posted the most incredible story – literally, the most incredible story that has ever been told since people have had the ability to tell stories, it will never, ever get as many hits as a video of a cat with a moustache' (Quah, 2015b).

In his article, 'Is this thing on?' (2014), producer Stan Alcorn looked into the bottlenecks which prevent audio from becoming a shareable medium. First, when you're listening to audio, you're usually driving, walking, doing the dishes. Rarely are you sitting at a computer. Second, audio isn't 'skimmable' in the way visual media and the written word are (Alcorn, 2014). Also, we would add, many people still prefer to find audio through a 'discover' feature on a podcatcher or by word of mouth.

We don't point this out to put you off the idea of posting your podcasts online. We point it out, first to manage expectations, but more importantly, to show how these difficulties can be turned into virtues. Social strategies for podcasts should combine a mix of community building with traditional methods of sharing. Unlike text and visual media, the goal isn't exponential growth. With the right strategy, your podcast can build a devoted community through off-(audio) platforms like Reddit, Tumblr or even Facebook. Though it's a slower process, it can be rewarding to create a space for your listeners to carry out the conversations you started in your podcast. It is worth repeating the Bertolt Brecht quote at the beginning of the chapter:

> The radio would be the finest possible communication apparatus in public life, a vast network of pipes. That is to say, it would be if it knew how to receive as well as to transmit, how to let the listener speak as well as hear, how to bring him into a relationship not isolating him. (Brecht, 1932: 15)

These are just a handful of encouraging success stories of people who have allowed their listeners to 'speak as well as hear':

- Nick Quah points out the success of the *Serial* podcast on social channels centred on the conversations it inspired online, although it did drive audience numbers through a grassroots following on places like Reddit. He writes, '*Serial* wasn't a podcast that went viral. It was the conversations around *Serial*, and certain derivative elements from the pod (like the Mailkimp meme), that went viral – propelling the existence of the podcast in front of greater and greater concentric circles of people' (Quah, 2015b). We recognise Reddit doesn't allow for self-promotion but the experience from *Serial* shows that a clear question, theme or central mystery (in other words, a story) encourages listeners to give feedback or seek out other listeners.
- *Welcome to Night Vale*, a fictional podcast, had a modest listenership until its fans made it an instant hit on Tumblr, just two years after its first episode. The majority of posts were of fan art of the fictional town. In July 2013, its downloads skyrocketed to 2.5 million as a direct result of the Tumblr exposure (McDonald, 2015).

And what of the traditional social media sharing that should run concomitant to these community-building strategies? As covered earlier, you can embed players from platforms like Soundcloud right into Twitter and Facebook. You can also create a page for your podcasts and garner 'likes' from interested followers, but Facebook is currently charging for the promotion of page posts, which can drastically limit your audience.

Another, more visual, approach is the creation of an audiogram. Quite simply, an audiogram is a snippet of audio on a player with a holding picture (a movie with a still photograph and sound). WNYC has an open source audiogram tool which automatically takes a portion of audio and makes it a video file for social media (Owen, 2016; see Figure 5.7). You can find the open source tool at: https://github.com/nypublicradio/audiogram. You can also make an audiogram easily with video-making tools like iMovie, Windows Movie Maker and Quicktime Pro.

Figure 5.7 Example audiogram from the WNYC podcast *There Goes the Neighborhood*

Credit: WNYC podcast There Goes the Neighborhood (www.wnyc.org/shows/neighborhood)

But for those organisations – such as international NGOs or universities – who already have huge Twitter and Facebook followings, do use these platforms to point your audiences towards your podcast content. You can schedule regular tweets to remind followers to subscribe on iTunes or share the podcast of the event they attended, or encourage more sharing by using podcast content as the foundation for competitions or give-aways. Some examples of effective tweets or Facebook updates could include:

- Our new podcast series investigates levels of corruption in international sport – listen here + Soundcloud link
- Don't miss our new podcast interview series on tackling gender pay gaps around the world + Soundcloud link
- Many of our followers have been in touch to say how much they enjoyed the new podcast – you can listen here too + Soundcloud link

Hashtags across all of these platforms are also an effective way to reach beyond your individual network. Be sure the hashtag is active enough to reach people but not so populated that your content will get lost in the noise. If the subject of your podcast is a trending topic that day, consider using the relevant hashtags to increase the chances of more eyes on your work. For example, if your podcast about your research on political activism in Brazil is published on the day of the opening ceremony of the Rio Olympic Games, use official and general hashtags about Rio and the Games to take advantage of increased public interest: #Rio2016 #Brazil #DilmaRousseff #Rio and #RoadtoRio, for example. More general hashtags about the subject of your podcast or your area of research can also be used: #activism #LatinAmerica #Brazil #favelas and #politics, for example.

An essential element of sharing anything on social media is learning what works. Measuring your efforts and keeping a record of the analytics, even if it is as basic as how many retweets or downloads your podcasts are receiving, will help you develop and refine your social strategy. Not only will this recording help you understand the reach and potential impact of your current work, it is also incredibly useful knowledge for your future projects and can save you time and stress down the line when it comes to facing these questions again.

5.5.3   Sharing podcasts through blogs

Great podcasts created with an interested audience in mind will most likely do well on audio platforms, with a bit of social media promotion to back them up. But an accompanying blog post that gives a short description of the podcast can add to the search engine optimisation of your work. A blog post is a social media-friendly method of dissemination – with blog platforms built for one-click sharing and easy social share counting and, to add SEO weight, with keywords most likely scattered throughout your accompanying write-up. When it comes to sharing a blog post about your podcast through social media, you'll ultimately be sharing the URL to the blog post, which will give readers a base to come back to and read more about your work and your project.

For an in-depth look at all things blogging – including choosing where to blog and how to blog – see Chapter 3.

5.5.4   Sharing podcasts in the 'real world'

Don't forget that the 'real world' exists too! After you've scripted, recorded, edited and published your podcast series, be prepared to discuss what you've created and how you did it at every opportunity. Conference presentations, public lectures and festivals are obvious venues to tell an engaging story about how you used podcasting at any stage of the Research Lifecycle, but there are plenty of other spaces. Many universities organise informal Show-and-Tells, TEDx talks and fringe-style events aimed at public audiences. Consider where your findings might provoke discussion and debate and look to see if there are any opportunities for engaging these audiences in dialogue.

Promoting your podcast series while networking in person can be of great help to your project. You can include the iTunes URL on your business cards or presentations, but avoid pressuring a new contact into listening on the spot!

5.5.5 Sharing podcasts through targeted emails

As discussed above, dedicated audio platforms offer the strongest route for sharing your audio content in an attractive and shareable format. But a combination of this plus social media promotion, 'real life' sharing and a targeted email campaign can really boost the reach of your podcasts too.

Though not as sexy as social media, email is still a considerable driver of web traffic and this is even more pronounced among mobile internet users. Figures from Campaign Monitor suggest that you are six times more likely to get a click through from an email campaign than you are from a tweet (Beashel, 2014) and for businesses email marketing has an ROI of 3800% (Direct Marketing Association, 2015).

Reach out to your colleagues via email and let them know about your new podcast series. Current colleagues, ex-colleagues, people you interviewed for the podcasts, their organisations, interested blog contributors – pretty much anybody that you've had related email contact with – should be on your list to email about your work.

Whatever method you pursue, make sure to coordinate a plan for sharing your podcasts. Focus on what specific audiences would be most interested in and would benefit the most from your audio content and look to build attention from there. In almost every case, an audience does not happen overnight. Continue to pursue opportunities where conversations, both online and offline, can occur and make sure to make notes on what works for the future. And if you are looking for more guidance, Chapter 7 includes more detail on choosing the right platform and planning a coherent social media strategy.

5.6 CONSIDERATIONS

By now, you should have a comprehensive overview of podcasting and what it can offer you as a researcher. We are in a unique stage of the development of podcasts, the medium is experiencing a resurgence, and suddenly advertisers and big-name media companies are heeding their huge potential. Through all the hype, though, podcasting, at its core, is a very old idea, as old as radio itself: a single voice, or maybe two or three voices, reaches the ear of a listener as they go about their day. But there is one key difference: that voice reaches the listener wherever they may be – on the way to work, on a hike or on a night's drive. Knowledge workers frustrated by journal publisher paywalls and low figure readership take note: podcasts bring your research to the wider world, mostly for free, and to an increasingly mobile consumer. It may not be an appropriate medium for every research project, but given the multitude of benefits it can offer an intrepid knowledge worker, it is certainly one worth seriously considering.

Here's a reminder of the key questions to ask yourself before you start podcasting:

- What do you want to achieve?
- What subjects and content do you have access to?
- Who are your target audiences?
- Is podcasting your research really necessary?
- What ethical considerations do you need to keep in mind?

There are a few legal points to consider when using sound or music from sources that are not Creative Commons. We encourage all podcast makers to be aware of the laws in their country. This section does not aim to give you comprehensive legal advice. If you have specific questions related to the use of licensed content, consult a legal professional. Most research organisations and universities can provide you with the name of a firm they work with regularly.

Fair use/fair dealing limits the exclusive right of copyright holders, allowing for other content makers to play or feature a portion of someone's work in their project. There are still restrictions on length of audio for purpose and character of the use. Generally, these laws allow for short, relevant segments of a collective work to be used for education purposes, criticism or review.

For example, in Britain you are allowed to use short extracts from TV shows and music but the law is still ambiguous on certain points. Songs can be thought to be comprised of several different elements: rights, music, the lyrics and the sound recording. To use the fair dealing exception under English copyright law, your work has to fit a number of clear copyright exceptions. One of these can be criticism or review, another is current events reporting. It is advisable to use only as much audio as you need, i.e. the portion of audio that you're referencing in your ensuing discussion, the clip that you want to reference and no more.

For more information on fair use in the US, visit the US Copyright Office's website: www.copyright.gov/title17/92chap1.html#107. In the UK, visit the UK Copyright Service website at: www.copyrightservice.co.uk/copyright/p01_uk_copyright_law. And in Australia, see the Copyright Council website at: www.copyright.org.au.

If this seems daunting, that's because it is. Copyright law is opaque. The good news is that when it comes to playing music, there are many places to find Creative Commons music should you want to score your podcasts. See: FreeMusicArchive.org, the Vimeo Music Store (https://vimeo.com/musicstore) and FreeSound.org.

5.7 FURTHER READING

Midroll Media (2015) 'The surprising secrets of successful podcasters.' Available at: www.
    midroll.com/free-whitepaper-surprising-secrets-successful-podcasters/
A comprehensive guide on how to start, produce and promote your podcast, pitched to
people at varying skill levels.

This American Life, www.thisamericanlife.org/about/make-radio
Advice and fun guides (including a comic book, videos and tip sheets) on how to make radio.

Transom, http://transom.org
Features great resources on the technical and practical means for podcasting and generally
creating great radio stories.

6

CREATING AND SHARING
PHOTOS AND VIDEOS
ON SOCIAL MEDIA

A great storyteller – whether a journalist or editor or filmmaker or curator – helps people figure out not only what matters in the world, but also why it matters. A great storyteller dances up the ladder of understanding, from information to knowledge to wisdom. Through symbol, metaphor, and association, the story-teller helps us interpret information, integrate it with our existing knowledge, and transmute that into wisdom.

– Popova (2014)

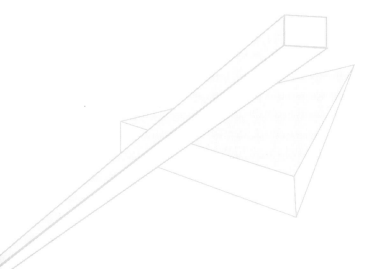

Figure 6.1 Michael Yorke and his wife, Valerie, with family members in the Ho tribe, Bihar, India, 1977

When Michael Yorke determined he would conduct research in rural India for his PhD fieldwork in 1972, he knew from the beginning he was going to make a film there. When his supervisor told him that the university would not accept a film as part of his submitted PhD, he continued anyway.

Yorke was going to study a tribe called the Ho in southern Bihar and he wanted a bigger and more engaged audience for his work than a PhD thesis or paywalled journal article could afford him. 'I wanted people to gain a human understanding of the Ho, which you can't get from an academic text,' said Yorke. The resulting film, *The Ho: The People of the Rice Pot* (1977), was one of the first observational documentaries of field work research and is now used internationally as a teaching film in anthropology (Figure 6.1).

Yorke recalls a 'magic' sequence in the film where a young girl walks through the forest on her own. The audience witnesses a closeness and transformation in her on that journey. Suddenly she opens up and becomes very candid about her relationship with her parents and family in this polygamist society. The scene was certainly of anthropological interest, but Yorke feels it also demonstrated the unique quality that film has for enhancing understanding of the human condition and telling stories. 'I think there are two aspects to making a film,' says Yorke. 'Reaching a wider audience is one of them, but the second is that it adds a more imaginative, passionate and emotional dimension to a textual analysis. It gives rise to a different understanding and interpretation of your research. I think film can be used in almost any scientific media' (personal correspondence, April 27 2016).

Yorke has spent the past 25 years producing films for British television, and has also spent several years at University College London as a senior tutor in ethnographic film. Today's aspiring research filmmakers have digital technology on their side: films are easier to make than ever before, and Yorke feels that 'in their films, the students have found expressive forms for their ideas and with these they have inspired and spoken to others in a fashion that the traditional essay has only very rarely done' (personal correspondence, April 12 2016). Today, of course, photos and videos remain as important as ever for telling stories from the world of research, and there are now more opportunities for sharing this content as photo- and video-driven social media platforms become ever more influential. Students and researchers are finding inspiration from online photo and video archives, as well as using social media to promote their own research films and ethnographic photographic work.

Whether you are a PhD student looking to learn more about how photos and videos on social media could feature in your work, a researcher looking to utilise video for your new blog and social media campaign, or a science communications professional wanting to share photos captured by astronauts in orbit, this chapter will show you how to make the most of photos and videos on social media. We will look first at some definitions of photos and videos on social media, then relate a brief history of their wider use in research communication, before examining the ways they can create impacts in a variety of settings. Finally, we will look at how you can create and adapt your own content using some of our favourite tools.

6.1 DEFINING PHOTOS AND VIDEOS ON SOCIAL MEDIA

Over the last 200 years, photos and videos have in some ways defined our social relationships and shaped how we conceive of reality. We have straightforward definitions of photography – such as Spencer's oft-cited description, 'the science, art and practice of creating durable images by recording light or other electromagnetic radiation, either electronically by means of an image sensor, or chemically by means of a light-sensitive material such as photographic film' (Spencer, 1973: 454) – but what about a broader definition that takes into account the societal and cultural impacts of these media?

The deeper question of how we define the act of creating photos and videos has attracted much attention. Bourdieu's *Photography: A Middle-brow Art* (1990) is one such influential work, exploring the social practice of 'taking pictures' and how there are few cultural activities more structural and systematic than photography. Encouraging us to question the belief that the photograph is an objective medium, that the camera never lies, Bourdieu argues that photographs offer us instead a reflection of the subject as we see it. Susan Sontag's *On Photography* (1977) is another important work on the topic, taking in capitalism, voyeurism and politics: 'To photograph is to appropriate the thing photographed. It means putting oneself into a certain relation to the world that feels like knowledge – and, therefore, like power' (Sontag, 1977: 2). Sontag's argument was made well before photo- and video-driven social media platforms existed, but still resonates today. The themes of power, appropriation and objectivity, raised particularly by Sontag, are still hugely relevant for us now: 'Needing to have our reality confirmed and experience enhanced by photographs is an aesthetic consumerism to which everyone is now addicted' (Sontag, 1977: 24). In the age of the social media photostream – 'the ultimate attempt to control, frame, and package our lives – our idealized lives – for presentation of others, and even to ourselves' (Popova, 2013) – *On Photography* remains a 'cultural classic of the most timeless kind' that can help us think about how we define photo- and video-driven social media platforms, and how the content we post there perhaps defines us too (Popova, 2013).

Throughout this chapter, we give explicit focus to digital photo and video content shared on social media platforms, rather than all types of photo and video, such as printed photographs. We have chosen to focus on both photos *and* videos on social media, and not split up the two forms into separate chapters. Some may feel that each merits its own chapter, or that podcasting and video may have more similarities, and we also acknowledge that there are ways in which photos and video differ. But for the sake of understanding how the content formats are used in conjunction with social media to communicate research, considering them together can be incredibly useful.

First, the overwhelming trend for social media content over the last few years has centred on photos and videos, and has privileged the visual over all other types of content. Daily communication and social interactions are becoming increasingly smartphone-based, meaning the sharing and consuming of photos and videos on social media platforms has evolved at a rapid pace. Every 60 seconds, more than 500 hours of video are published on video-sharing platform YouTube; every day, more than 80 million photos are published by users on video- and photo-sharing app Instagram; and the photo, video and messaging app Snapchat now has over 10 billion daily video views (Smith, 2016). We can think of social media platforms as starting their life course as predominantly text-based, then progressing to experiment with photos and videos, and now, from 2017, we are into the world of live video and augmented reality apps.

Second, for those who are using photos or videos on social media to share their research, we often find that their aims are similar: to achieve a deeper emotional connection with research audiences and to provide a rich storytelling experience. By choosing to use photos or videos on social media platforms instead of other content, creators are often seeking to achieve a more human, personal touch that their audience can connect with, at whatever stage of the Research Lifecycle. We will cover this in more depth in section 6.3

when we look at the ways photos and videos on social media can contribute to impactful interactions. But first, we will discuss the historical development of using photos and videos to communicate research.

6.2   PHOTOS AND VIDEOS: A HISTORY IN RESEARCH COMMUNICATION

The collective history of photography and videography is a fascinating one, taking in the first pinhole cameras created by Persian philosopher Ibn al-Haytham over 2000 years ago; the first commercially successful photographic process of daguerreotypes from 1839; more recently, to the iconic Super8 of the 1960s and the bulky home movie cameras of the 1980s; and right up to the mobile photo and video technologies on our smartphones and tablets that we are familiar with today.

The full history is not ours to cover in a chapter that focuses on the use of photos and videos on social media. However, it is worth considering how these media forms have been used through recent history for research communication and how they connect to social media interactions today. Though these forms of media may at first appear to be an unfamiliar and daunting area for some academics and researchers, this chapter shows that – just like blogs, data visualisations and podcasts – photos and videos have been used innovatively for decades across many areas of research communication.

6.2.1   Social documentary photography: From Victorian London to going viral on Facebook

It is true today, but particularly so during the Victorian era, that social documentary photography has been used 'to expose evil and promote change' (Becker, 1995: 7), allowing us to hold up a mirror to society 'by focusing on painful facts and little-known people in hopes of making beneficial changes' (Street, 2006: 386). This style of photography is built around a core value that many working in research and academia will appreciate: the desire to make a positive political and social impact.

An early example of social documentary photography was undertaken among the stinking London slums during the Industrial Revolution, with sociologist-journalist Henry Mayhew and photographer Richard Beard recording the brutal reality of life for the working classes in *London Labour and the London Poor* (1861). Described by respected historian Peter Ackroyd as 'a remarkable and affecting source of street life' (2001: 142), the interviews with and images of working-class Londoners revealed the grim reality of finding employment as rat catchers and sewer hunters; else face desperate hunger in disease-ridden slums. Mayhew felt that in order to give the interviews factual credibility

(Groth, 2012), it would be necessary to use accompanying photographs of his interviewees, noting that 'until it is seen and heard we have no sense of the scramble that is going on throughout London for a living' (Mayhew, 1861: 10).

In the decades that followed, photographer John Thomson and radical journalist Adolphe Smith made a further case for social reform, with their *Street Life in London* (1876), which is now regarded as a key work in the history of documentary photography. Historian Roy Flukinger writes of Thomson's 'serious commitment to his subjects and his art … his compassion' (1985: 83) (see Figure 6.2). We also see similar themes in the work of social documentary photographer Jacob Riis, who was working to the same ends in New York in the 1880s (see Figure 6.3). All of these works contributed to an increase in public pressure on governments to help the most vulnerable in society.

Figure 6.2   *The Crawlers*

Credit: From John Thomson and Adolphe Smith's *Street Life in London* (1876)

Figure 6.3 *Bandit's Roost*, a Mulberry Street back alley, photographed by Jacob Riis in 1888, a target of police efforts in the 1880s and 1890s

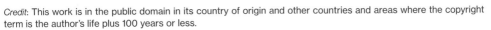

Credit: This work is in the public domain in its country of origin and other countries and areas where the copyright term is the author's life plus 100 years or less.

Looking at the historical connections between Victorian-era social documentary photography and the photos and videos shared on social media today, we can see their influence in a variety of popular social media and blogs, including *Humans of New York* – the hugely admired blog and bestselling book project run by photographer Brandon Stanton, which has over 17 million Facebook followers (www.facebook.com/humansofnewyork) as of August 2016. Stanton started by documenting everyday people on the streets of New York on his Facebook page: taking a natural portrait photo and interviewing them about their lives and what advice they might like to share with the world. Within months the profile of *Humans of New York* had grown internationally and thousands of Facebook followers were sharing the compassionate and considered photos and interviews. As the popularity of his Facebook project has grown, Stanton has developed photo series that capture the stories of people living in Iran, Iraq and Pakistan, and he has even travelled to Europe with UNHCR to photograph and interview refugees travelling across oceans to flee violence in the Middle East. His work now more broadly aims to challenge negative stereotypes in the media or wider society, and he frequently teams up with humanitarian organisations.

Many of Stanton's images go viral, with tens of thousands of shares and comments. Stanton often does more than sit back and watch the networked nature of social media bring in the shares and comments, though. His photos – capturing with precision emotions of sorrow, joy or love the way that a written essay so rarely can – are used as the starting point for fundraising and philanthropy. A fundraising campaign in 2015 raised more than $1 million in donations for visits to universities for school children, after Stanton photographed and interviewed 14-year-old Brooklyn resident Vidal, who shared how his inspiration was his teacher, Mrs Lopez. As a result of the campaign, Stanton, Mrs Lopez and Vidal were even invited to visit the White House, and their story and campaign made headlines around the world.

Just like the very early social documentary photographers, it is the compassion and honesty that comes through in these photos that makes them so remarkable, and elevates this medium above the written word.

6.2.2 Dorothea Lange's Dust Bowl photographs define an era and inspire today's Pinterest users

Eighty years after Mayhew, sociologist and photographer Dorothea Lange captured images that defined Depression-era America and were part of a project that redefined how photos are used in social research today.

When huge swathes of once fertile prairie farmland in the US and Canada were devastated by severe dust storms, tens of thousands of families migrated to California and other states for work. They soon found that the Great Depression had rendered economic conditions there little better than those they left behind. More Dust Bowl families continued to arrive, with no work, no pay and little shelter. The US Department of Agriculture's Farm Security Administration (FSA) tasked photographers, including Lange, to document the crisis, and in turn promote the FSA resettlement campaigns.

It was here, among the dried-out farmhouses and dusty fields, that Lange took some of the most iconic photographs of the 20th century – including *Migrant Mother* (1936) (see Figure 6.4). The photo shows Florence Owens Thompson with two of her seven children, waiting at a pea-pickers' camp in Nipomo Mesa, where hopes of work and pay were dashed after freezing rain destroyed the crops.

The photographs prompted a public outcry after being published in the press. Lange's images succeeded in gaining political impacts for the FSA, solidifying the belief of project economist Paul Schuster Taylor that 'words alone could not convey' the reality of life for farm workers, and he wanted his department's reports to have the maximum possible impact (Whiston Spirn, 2008: 17).

Lange's photographs are historically significant and 'live in the subconscious of virtually anyone in the United States who has any concept of that economic disaster' (Gordon, 2006: 698). With its Madonna and child imagery, *Migrant Mother* became 'a visual reference point for a vast amount of photographic coverage of the world's harsher events'

(Jacobson, 2001: 1454), and its influence can be seen in photographic portrayals of other crises, including the Rwandan genocide, Hurricane Katrina and the European migrant crisis of the mid-2010s. The FSA photography project closed in 1942 with an archive of more than 160,000 photographs and is widely recognised as 'the best example to date of the use of photography as a tool for social research' (Meadows, 2015: 69).

Figure 6.4 *Migrant Mother*

Credit: Dorothea Lange (1936). This image is available from the United States Library of Congress's Prints and Photographs division under the digital ID fsa.8b29516

There is a strong connection between Lange's photographs as era-defining objects and tools for social research (see, for instance, Figure 6.5), and the content curation element of many photo- and video-driven social media platforms that are popular today. Writing for the *LSE Impact Blog*, Amy Mollett (2013) discusses how Pinterest can be used by researchers:

> Pinterest has a reputation for the over representation of cupcake recipes and cute crea-tures amongst its pins, but there's no reason why it can't be used to create boards centred around learning and research. With Pinterest, users pin relevant images to their themed boards – in this case we're talking about book covers, author images, archive footage, TED Talks, etc. Pins can be re-pinned from another user's board, or taken from websites, and users also have the option of uploading images from a computer. (Mollett, 2013)

The Sociological Cinema Pinterest feed (https://uk.pinterest.com/thesocycinema) is one such example and promotes the wider *Sociological Cinema* initiative run by three public sociologists at the University of Maryland. Their website provides resources for teaching and learning through video, but their Pinterest boards are packed with over 9,000 pins and have over 10,000 followers as of mid-2016. *The Sociological Cinema* team uses Pinterest to curate images that define moments or eras in society and culture – just like Lange's – and have pinboards dedicated to topics such as disability, street harassment, race and ethnicity, and others for social movements like Black Lives Matter, the March on Washington for Jobs and Freedom, and most recently the protests following the election of Donald Trump as President in November 2016 (Figure 6.6).

Figure 6.5 On Arizona Highway 87

Note: Lange's description of the photogrpah: On Arizona Highway 87, south of Chandler. Maricopa County, Arizona. Children in a democracy. A migratory family living in a trailer in an open field. No sanitation, no water. They came from Amarillo, Texas. Pulled bolls near Amarillo, picked cotton near Roswell, New Mexico, and in Arizona. Plan to return to Amarillo at close of cotton picking season for work on WPA.

Credit: Dorothea Lange (1940) for Department of Agriculture. This media is available in the holdings of the National Archives and Records Administration, cataloged under the ARC Identifier (National Archives Identifier) 522528

 Lester Andrist, part of the team behind *The Sociological Cinema* Pinterest feed, shared his experience with us:

> As with all the social media we use, the impetus to use Pinterest was born from a desire to reach more people. From very early on, it became apparent that Pinterest offered some distinct advantages over other social media platforms. First, we loved the way our main profile page showcased the boards we created. Since *The Sociological Cinema* is

interested in promoting the teaching and learning of sociology, it made sense to roughly base these boards on the subfields of the discipline, such as race, social class, gender, social movements, etc. We wanted to populate each board with images, quotes, and figures that pertained to each of these topics. ... The idea is that students can open the boards that align with their interests and find themselves immersed in information and provocative images. It's a place to explore a topic and possibly find ideas for research projects. Sociology instructors have also told us that they include the boards as free supplementary material in the classes they teach (personal correspondence, April 12 2016).

Figure 6.6 The Sociological Cinema Pinterest board

Source: https://uk.pinterest.com/thesocycinema/

Visual content – whether it is Lange's original archive of photos or Pinterest pins of iconic imagery from social movements – can be useful in helping to illuminate histories. Curating a collection of images that define key social and cultural moments can be a very effective way to present this information. With the growing popularity of photo- and video-driven social media platforms, researchers can now pull together useful resources, engage wider audiences through relevant themes and topics, and link this information to wider social action.

6.2.3   Photos and videos in science communication: Iconic NASA moon landings to Instagramming astronauts

It is not just social reformers in the social sciences who have made historic uses of photos and videos. Scientists at NASA have been creating and sharing innovative photos and videos for educational purposes since the 1950s, and they continue to work magic, innovating on photo- and video-driven social media platforms today. In practice, this has meant everything from live international TV broadcasts during mission launches, to making all NASA-produced photos available to use in the public domain, to having a wildly active social media presence that allows students to interact directly with astronauts on the Space Station.

Right from its inception, a core value of NASA was to communicate its activities and information to the world as widely as possible. Its founding legislation set out that the Agency was to 'provide for the widest practicable and appropriate dissemination of information concerning its activities and the results thereof' (NASA, 1958). Throughout NASA's six-decade history, photos and videos have been at the heart of these communication efforts. The photos and video footage of NASA's Apollo 11 – the space flight that landed humans on the moon – comprise the first instance that truly brought to life NASA's aim of maximising public interest in space, showcasing American space expertise and power in this era. As Apollo 11 left on July 20, 1969, over 600 million viewers watched the film footage on their TV sets worldwide to see the moon landing. Those now iconic images defined an era: the space race was over and a new epoch was upon us (see Figures 6.7 and 6.8).

Figure 6.7 Vice President Spiro Agnew and former President Lyndon B. Johnson view the liftoff of Apollo 11 from pad 39A at Kennedy Space Center at 9:32 am EDT on July 16, 1969

Credit: NASA On The Commons. www.flickr.com/photos/nasacommons/9460201254/in/album-72157634973926806/ Image # : 107-KSC-69PC-379

Just as the first mission to the moon heralded some of the most important photographs the world had ever seen, so too did the final moon mission: Apollo 17, on 19 December 1972. The crew had taken a photograph that would go on to become one of the most – if not the most – reproduced photographs in human history (Reinert, 2011; Riley, 2012). The first photograph of Earth as a fully illuminated whole, it was nicknamed the 'Blue Marble' and was taken at a distance of about 45,000 km (28,000 miles) from Earth (see Figure 6.9). Printed on the front page of nearly every newspaper in the world, its beauty and the stories it told captivated countless readers. Published – just like all other NASA images – in the public domain, it was embraced by green thinkers and peace activists: 'everyone from NGOs working in the developing world to the environmental movements seeking to protect our planet: For 40 years it has been used to change minds, behaviours and political policies' (Riley, 2012).

Today, NASA shares photos and videos through more than ten social media platforms in its educational and engagement efforts, including on Facebook, Flickr, Twitter, YouTube and Instagram. In 2009, NASA astronauts sent the first tweet from space (see Figure 6.10); in 2010, they did the first Foursquare check-in from space; in 2011, they were the first US Government Agency on Google+; and in 2012, they had the first Foursquare Check-in from Mars (NASA, 2012).

Figure 6.8 Aldrin on the moon

Note: Astronaut Edwin E. Aldrin, Jr., Lunar Module pilot, is photographed during the Apollo 11 extravehicular activity (EVA) on the lunar surface. In the right background is the Lunar Module "Eagle." On Aldrin's right is the Solar Wind Composition (SWC) experiment already deployed.

Credit: Taken by Neil A. Armstrong with a 70mm lunar surface camera. NASA Image #: AS11-40-5873

Figure 6.9 The 'blue marble' photo of the Earth taken from Apollo 17

Credit: NASA Goddard Space Flight Center Image by Reto Stöckli (land surface, shallow water, clouds). Enhancements by Robert Simmon (ocean color, compositing, 3D globes, animation). Data and technical support: MODIS Land Group; MODIS Science Data Support Team; MODIS Atmosphere Group; MODIS Ocean Group Additional data: USGS EROS Data Center (topography); USGS Terrestrial Remote Sensing Flagstaff Field Center (Antarctica); Defense Meteorological Satellite Program (city lights).

Figure 6.10 Mike Massimino's first tweet from space

Credit: https://twitter.com/astro_mike/status/1777093627

NASA has many millions of followers across hundreds of accounts and scores of platforms. One of the reasons for its success is down to the fact that the stunning photos and videos of the Aurora Borealis, the International Space Station and the sun rising over Earth are taken by the people who are experiencing those images first-hand: the astronauts. The fact that the astronauts taking these images are the only human beings able to do so allows us mere Earth-bound followers to experience the beauty and awe of space first-hand: 'breathtaking, inspiring – and perfect for Instagram' (Vazquez, 2015) (see Figures 6.11 and 6.12).

Figure 6.11   Instagram from Scott Kelly @stationcdrkelly

Credit: www.instagram.com/p/-MSE25gXoL/?hl=en

Figure 6.12   Instagram from Scott Kelly @stationcdrkelly

Credit: www.instagram.com/p/9816D7AXID/?hl=en

NASA has proven time and again that its historic photos and videos are a stand-out way of generating public and media interest in its activities. The approach is a convincing example of how photos and videos can play a leading role in research communication and how social media can work as a prime conduit for these messages.

6.3 WHY PHOTOS AND VIDEOS ON SOCIAL MEDIA ARE USEFUL ACROSS THE RESEARCH LIFECYCLE

Today, the growth of photo- and video-driven social media platforms is enormous, with Instagram, Snapchat and Pinterest breaking records every few months for numbers of users and amounts of photos added each day. As of mid-2016, Instagram has over 300 million monthly active users, YouTube has over a billion – that's one-third of all people on the internet – and Facebook generates 8 billion video views per day.

We also know that social media audiences find this content most engaging. On Facebook, photos are the most engaging type of content with a whopping 87 per cent interaction rate from fans, and account for 75 per cent of content posted by Facebook pages worldwide (Ross, 2014). Between April 2015 and November 2015, the amount of average daily video views on Facebook doubled from 4 billion video views per day to 8 billion (Constine, 2015). Research that analysed the content of over 2 million tweets sent by thousands of Twitter users over the course of a month showed that adding a photo to your tweet can boost retweets by an impressive 35 per cent, and adding a video can boost retweets by 28 per cent (Rogers, 2014). Clearly there are opportunities here for researchers to infiltrate these spaces, get wider exposure for research messages, solicit feedback and engage with new audiences.

But in an era where the impact of research matters more than ever and when social media provide us with opportunities to reach ever-expanding audiences, we need to understand more about what impacts using photos and videos can have. How do we know whether the research findings in our videos are reaching and are even being understood by the intended audiences? Are photo- and video-driven social media platforms really worth using to support impact and knowledge exchange goals, or are they just a fad? We will now explore some of the impacts behind research projects that have used photos and videos on social media to disseminate their work, at a variety of places in the Research Lifecycle.

6.3.1 Photos and videos on social media can help academic discussions become more accessible and useful to the public

Mark Blyth, a professor of international political economy at Brown University, became a YouTube sensation in 2011 after hosting a five-minute video criticising the so-called 'common sense' of austerity policies being introduced by governments across the world. His video made clear the connections between class politics, media rhetoric and the financial crisis.

Many economists, political scientists and sociologists had been critical of the politics of austerity for months but little of this criticism was breaking through into mainstream discourse. Blyth already had experience with public engagement activities through blogging about his research when Brown University matched him with videographer Joe Posner to create a video on austerity, as the topic was beginning to make headlines in the UK for the severity of cuts being made. The slick video consists of attractive and engaging visuals, with Blyth speaking to the camera in a relaxed fashion and in accessible language, while Posner's eye-catching animations aid our understanding of Blyth's arguments and keep the viewer focused. You can watch the video in full at youtube/go2bVGi0ReE.

For Blyth – and for many of his viewers – the video is a fantastic example of how to take a fairly complex topic and make it understandable. Blyth sums this up nicely in an interview he did for the *LSE Impact Blog* in 2012:

> The two core ideas of the video – that you can't cure debt with more debt, and that while any one state can cut its way to prosperity it can't work if they all do it at the same time – are technically known as 'trying to fix a solvency problem with a liquidity instrument' and 'a fallacy of composition'. Start by posing the problem in that language and it's dead to anyone who is not an expert in the debate already. Turn it into things people can understand, let go of the academese, and people will engage. The video was posted first on YouTube and was then massively reposted by Facebook. I got fan mail and hate mail from all over the world. If you want impact, this is how to do it. The old days of doing an edited volume with your mates and publishing it a year and half later as 'impact' is increasingly a dead model in an era of social media.

The video was a huge success and was written about on charity hubs, international political blogs and in newspapers. It racked up tens of thousands of views in just a few months, and as of mid-2016 it has been watched by over 100,000 people across YouTube and Vimeo. Blyth was then asked to write a book based on the video.

For Blyth, the video had three main impacts. He explains:

> First, it raised my profile globally, which encouraged me to do more video as a way to communicate ideas. My most recent video for AthensLive, predicting 'Brexit as Trumpism' in early June 2016, for example, had over 850k Facebook shares and over a million views off platform. That gets noticed. Second, it convinced me that there is a real demand for academic engagement with the public debate, so long as you can 'drop the academese' and explain your ideas clearly. When you can do it with video, especially animation, that only amplifies the message. Third, it changed what I do as an academic. I now think about the visual impact of ideas as a core goal of my research. (Personal Correspondence, August 1st 2016)

In terms of advice for others considering using video to promote their research, Blyth recommends finding an animator or working with your university or institution's video team: 'They are generally speaking very keen to do this type of work and you only have to ask.' Find a theme 'that you care about and want others to care about, and then strip it down to why it matters and what people need to know. Practise it, storyboard and script it. Shoot it and let it out there. Tweet the heck out of it. Sit back and enjoy the show.'

Of course, not all videos that make complex subjects understandable will go viral, but in this case, a formula of professional production, some established channels for sharing, a charismatic host and a topic that captured the *zeitgeist* helped to build impact or an audience around the project. Equally, there are cases where high-production value isn't necessary. YouTube has in many ways opened the doors for researchers to turn on a camera and talk about their subjects in approachable ways. For example, sociology professor Simon Lindgren has had enormous success with his self-produced Social Science in 60 Seconds series on YouTube (www.youtube.com/user/simonlindgren), which has to date received over 60,000 views. But cutting through the noise of YouTube can be difficult. As the austerity video shows, weaving a strong narrative with the addition of visual animations can certainly help to drive your points home.

Would a blog post or podcast have done the same job? Perhaps, but the visual effects that reinforce Blyth's arguments are a big part of the video's success. Would a journal article sitting behind a paywall have had the same level of interest? Most likely not. By actively choosing a medium that everyone can access, and by using a conversational and open style, Blyth's austerity video is one of the most exciting examples of videos to have sparked wider public engagement in politics and economics in recent years.

6.3.2 Photos and videos can challenge entrenched researcher–participant barriers

Digital photos and video can be used effectively at any stage of the Research Lifecycle, not just the dissemination stage. A perfect example of this is in the primary research phase. Barriers between researchers and participants are well documented (Denzin and Lincoln, 2005), but using photo-elicitation techniques – where photos, videos or any type of visual image are used in an interview to start discussions with participants – can promote 'more direct involvement of the informants in the research process' and 'encourage and stimulate the collection of quantitatively and qualitatively different information to that obtained in conventional interviews' (Bignante, 2010: 2). Douglas Harper's (2002: 13) history of the development of photo-elicitation in anthropology and sociology outlines why it is that images can be a powerful force in triangulation between different information sources:

> The difference between interviews using images and text, and interviews using words alone lies in the ways we respond to these two forms of symbolic representation. This has a physical basis: the parts of the brain that process visual information are evolutionarily

older than the parts that process verbal information. Thus images evoke deeper elements of human consciousness that [sic] do words; exchanges based on words alone utilize less of the brain's capacity than do exchanges in which the brain is processing images as well as words.

One recent example is the 'Gender and Violence in Urban Pakistan' project run by Dr Amiera Sawas as part of the Safe and Inclusive Cities Programme (Anwar et al., 2016). The aim of the project was to identify the drivers of violence in urban spaces through photography. Sawas had long been interested in participatory photography but had yet to see it executed effectively by an academic, which motivated her to try it herself. Curious, she sent out a notice through a LinkedIn group devoted to the methodology and received a lot of responses from researchers and photographers who had done similar projects to the one she wanted to carry out in Pakistan. From there, she collated advice on the types of digital camera and the best ways to carry out a project on a budget.

Participants were given a digital camera and asked to take photos of their lives and surroundings for four weeks (see, for instance, Figure 6.13). Key themes that participants were asked to focus on were fears, comfort, sadness and happiness. Sawas felt the photo-elicitation method captured something the questionnaires could not: 'I think it was the most authentic conveyance of their feelings because we weren't guiding them' (personal correspondence, April 19 2016).

Sawas and her team are applying for further funding to share their findings in a gallery viewing in Karachi with the goal to challenge middle-class assumptions about other social classes – something that is not easy to do with a journal article. Conventional forms of research dissemination don't have the same impact. Sawas again: 'If the research we are producing is relevant to society, why aren't we communicating it? If we aren't interested in sharing and starting conversations, then I don't see the point of doing what we do' (personal correspondence, April 19 2016).

Figure 6.13 'We have ways to construct a door to our washroom but we don't have money,' writes Fatima Ali from Karachi, 'It's really painful being a girl and using this washroom'

Credit: Photos taken by participants in Sawas's project

Digital photos and video lend themselves well to the dissemination phase of the Research Lifecycle, but incorporating them at other stages can produce great results too. As mentioned in Chapter 2, live feeds of field sites are a growing phenomenon in crowd-sourced archaeological projects. Particularly for social research, photo and video formats can reshape the restrictive dynamics between research and participant, and allow partici-pants to have their say in defining what is researched and how. But photos and videos are not without their challenges in this environment. New barriers can be erected when it comes to unfamiliar technology and it isn't hard to imagine scenarios where integrating video recording actually hinders wider involvement in the research process. But given the more human-centred perspectives that can be gained and the conversations that this can open up, these formats are certainly worth experimenting with further in the primary research phase.

6.3.3 Photos and videos on social media can extend the life, value and impact of your research

Having an online home for your project's photos and videos means that the life and accessibility of your work are immediately extended. Rather than only having one public event about your new research project – over and done with in the space of two hours and limited only to a few hundred attendees – using photos and videos on social media to make a record of the lecture and sharing this online mean your work has the potential to be found by a much larger audience.

As government funding for the arts becomes continually scarce in the UK and around the world, many arts organisations are under pressure to show value for money and make clear how the public are able to access as much of their collection as possible. Using video to livestream or record tours or lectures at galleries for social media platforms means that funders get increased value from their investment: the knowledge shared will continue to live on and be pushed out to a wide audience, while the institution itself engages in an important knowledge exchange activity.

Nottingham Contemporary art gallery is one pioneer in the arts for adding video lecture content to its YouTube channel, social media feeds and website. An organisation with a strong emphasis on promoting the exchange and dissemination of ideas inspired by today's art practices across disciplines and cultures, it replicates this approach through its online video offerings. The gallery records videos of its public events with innovative designers and leaders from the contemporary art world. By sharing these on its YouTube channel, thousands of people have been able to watch and learn. Its two most popular vid-eoed events are from renowned comic creators Alan Moore and Melinda Gebbie, reflecting upon the relationship between art, underground publishing and radical politics (https://youtube/93sV5XGLmgQ), and from Professor Stuart Elden on Foucault, Subjectivity and

Truth (www.youtube.com/watch?v=abdLEdnZ_kQ). Each video has received over 25,000 views – many more than the 200 people in the audience on each night.

In another example, the London School of Economics is increasingly using video lectures via its YouTube channel to open up its public lecture programme as part of its knowledge exchange strategy. Livestreams allow social media followers to get involved with discussions via the hashtag, showing how video can link up well with your project's other social media and digital offerings. On average, these videos receive 5,000 views on YouTube – 15 times the figure for in-person attendance.

2016 saw LSE also trial a new approach to showcasing academic opinions: a series of 60-second videos featuring a variety of academics discussing key issues surrounding Brexit. The #LSEBrexitVote video series was launched in mid-April, ten weeks out from the EU referendum on 23 June 2016. The videos posted on LSE's Facebook page alone attracted 259,000 views, with the accompanying tweets making 611,348 total impressions and 4,584 engagements. The #LSEBrexitVote YouTube playlist attracted 19,383 views – ten times the average views for an LSE playlist – making the series the highest ever watched playlist in the School's social media history.

This video experiment was a first for LSE, combining academic expertise with a strong news angle, public interest and visual appeal, and packaged to a social media-friendly 60-second run time. The videos were promoted by the Media Relations team to national and international journalists who went on to browse the videos for potential follow-up interviews. Candy Gibson, senior press officer at LSE, sums up the impact for the institution in a 2016 blog post:

> Not only did the project showcase the most important issue facing the UK for decades, but it also aligned it with the School's strategic priorities – demonstrating strong research, encouraging diverse views, influencing government policy, providing solutions to key social challenges, cementing the LSE brand as a world leader in politics and social sciences, and demonstrating LSE's intellectual engagement with the wider world.

Joining the dots between politics and the arts, the Center for Public Scholarship at The New School recorded and posted to YouTube its public conference on 'The Fear of Art': artists, scholars, museum directors and audiences discussed the power of art and the importance of advocating for art, artists and freedom of expression. Ai Weiwei gave the keynote address via a livestream from China, on 'The Censorship of Artists: Artists in Prison, Artists in Exile' (www.youtube.com/watch?v=YnBnjcKYuRs), reaching thousands of viewers and attendees internationally via livestreamed video and social media activity. The video – like many livestreamed offerings – has been made available for viewers to watch on demand, anytime, on YouTube.

As well as universities and those in Higher Education, other organisations that centre on knowledge have also heartily embraced the platform. Instagram has enabled museums and art galleries such as the Museum of Modern Art (@themuseumofmodernart), the Louvre (@MuseeLouvre) and the Tate group (@tategallery) to communicate the visitor experience through social photo sharing. Snapping not just photos of paintings, sculptures and installations, but also focusing on how visitors interact, study and find inspiration in these spaces is at the heart of their Instagram strategies.

Photos and videos shared and livestreamed via social media today are acting as a record of research's influence in the world. Wider engagement through social media channels and the repurposing of existing content streams have direct bearing on how research activities are aligned with the strategic objectives and civic missions of organisations.

6.3.4 Photos and videos on social media can help charities raise money and awareness

As well as academic research projects, NGOs and charities have also had huge successes in using photo- and video-driven social media platforms to execute international campaigns and get their followers actively involved in support and fundraising. Remember the Ice Bucket Challenge of summer 2014, which helped raise awareness and funds for amyotrophic lateral sclerosis (ALS)? More than 17 million people uploaded videos challenging their friends to dump a bucket of icy cold water over their heads (see Figure 6.14). These videos were watched by 440 million people a total of 10 billion times, raising over $100 million in one summer. Using video to encourage a competitive spirit, with a sprinkle of narcissistic social media pressure and low barrier entry so that anyone could make a video on their smartphone, the Ice Bucket Challenge is considered by many commentators to be a campaign that defines a new era of video use on social media. In June 2016 the project's researchers announced that they had identified a new gene associated with the disease which could lead to new treatment possibilities – all done with the funds raised through the Ice Bucket Challenge.

Not every organisation is going to be able to raise as much money and awareness as the ALS Ice Bucket Challenge, but there is growing agreement among fundraising professionals that photo- and video-driven platforms are one of the best tools 'for storytelling and connecting supporters emotionally with nonprofits' work' (Amar, 2015). In an interview with *The Guardian*, Nick Owen of Médecins Sans Frontières shared that the @doctorswithoutborders Instagram feed tries to 'bring our supporters as close to MSF's frontline work as possible. Whether it's trauma surgery in a conflict setting or mobile outreach care in Ebola hotspots, Instagram is a fantastic way to give an up-to-date insight into our lifesaving work in over 60 countries. It's opened us up to a wider, and younger, audience' (Amar, 2015).

Figure 6.14 Anthony Quintano completes the ALS Ice Bucket Challenge

The social nature of platforms like Instagram also allows for new levels of interactions with followers and supporters. The National Trust @nationaltrust runs a photo challenge every weekend asking its Instagram community to share photos of their favourite properties and gardens, using the hashtag #NTChallenge. This boosts engagement and enables the charity to share the stories behind its properties in a way that just wouldn't be engaging in a written format.

Pinterest has been a great match for many charities and NGOs due to female donors representing a growing force in the world of online giving. Over 60 per cent of all Pinterest users are female, so it makes sense for charities to have a lively presence on this social network. Though the days of overspending on printed marketing materials may not be over, nonprofits are able to save money by effectively utilising their social media channels, and Pinterest gives us some great examples of that.

The NGO Skateistan began as a grassroots 'Sport for Development' project on the streets of Kabul in 2007 and is now an award-winning, international NGO with projects in Afghanistan, Cambodia and South Africa (see Figure 6.15). Working with children aged 5–18, and with 40 per cent of participants girls, Skateistan is the first international development initiative to use skateboarding as a tool for empowering youth, creating new opportunities and the potential for change. The work it does through skate schools and community building makes a considerable difference in individual participants, but the organisation has also been instrumental in challenging dominant media representations of the country and its residents. Skateistan's strong social media presence – across seven different platforms and with hundreds of thousands of international followers – is dominated by bright, eye-catching photos and videos of activities at the schools shared on Instagram, Facebook and YouTube.

Reports and journal articles about the project may well be read widely, but they are unlikely to have the same immediate and accessible messages in them about the positive work being undertaken by the NGO that its social image sharing is able to convey.

Figure 6.15   Skateistan participants in Afghanistan

Credit: www.instagram.com/p/80TJ5uKDqE

6.4   HOW TO CREATE SUCCESSFUL PHOTOS AND VIDEOS FOR SOCIAL MEDIA

Up to this point we've covered some of the major events and trends in the history of photos and videos for research communication, and we have also explored the impact of their use in a variety of settings. It is clear in each of the examples that visual storytelling through social media is an effective strategy for communicating research. So how can you get started with this for your own research project?

We start by exploring the questions you should ask yourself and your team before you undertake this type of project; then we look at different examples of photo and video types and which ones might work best for your project, wrapping up with some technical and legal considerations.

6.4.1   The questions to ask before you create anything

Before creating and sharing any photo or video content for your project, it is essential that you take the time to consider some key questions around what you want out of your project and the audiences you intend to reach. This will not only allow you to reflect critically on the means by which you will disseminate your content, it will also allow you to address any potential barriers before they risk becoming a hindrance later on.

Given that photos and videos on social media are in the same multimedia category as podcasts, you may recognise some similarities between these initial stage inquiries and the ones that we have laid out in the podcasting chapter.

The questions to ask before you create anything:
1   What do you want to achieve?
2   What subjects and content do you have access to?
3   What story do you want to tell?
4   Who are your target audience and what are your aims?

Question 1: What do you want to achieve?

Why do you want to create photo or video content for social media? Is it because you want to reach your audience in an eye-catching way? Or are you trying to make a policy impact? Consider whether the images and footage you intend to gather will have the effect or impact you want, and whether your audience will be interested enough to engage with it. Videos are great fun but they are not always appropriate for political impacts, for example. Setting out your aims and objectives will focus your efforts before your head gets wrapped in more technical tasks.

Question 2: What subjects and content do you have access to?

If you are in the primary research phase, you should be thinking about the participants and partners you might have access to and whether they would be happy to be interviewed, captured in footage, and have their images shared through your social media feeds. In many cases, consent from the individuals will need to be obtained in order to use the images fairly and lawfully. Data protection, digital storage for your material and photographer–subject power imbalances should all be taken seriously.

In terms of the dissemination and impact phases of the Research Lifecycle, think about any events attached to your project: could photos and video footage from lectures or pop-ups be curated and shared through your social media feeds? Or could you create an animated video that uses elements of other things you might already have created, such as blog posts, data visualisations or podcasts?

If you're not working directly with any subjects (animal, human or other) or you'd like to experiment a bit first, there are several free photo and video archives online that you can explore and source content from. Whatever you are working with, planning out what you could capture and where you could capture it will help you produce better results. It will also ensure that you obtain the right permissions before committing fully to a particular research subject, should you still be in the primary research phase.

Question 3: What story do you want to tell?

When thinking about how you might disseminate your research using photos or videos on social media, explore what specific stories you can tell. As you're sharing your content on social media, you need a punchy, clear delivery. Think of a story in the classic sense here: exposition, rising action, climax, falling action and resolution.

For example, for somebody with a journal article about the findings of a study on internet safety for teenagers who is looking to create a one-minute video aimed at school teachers, we can use these simplified steps. First, lay out a problem or phenomenon – many children and teenagers use social networks today, but there are dangers to be aware of. Second, identify a potential solution for that problem or, if it's a phenomenon, provide a potential explanation – the results of your study show that talking to children about the subject in a way that empowers them and gives them specific action points or routes to find help is the best way to tackle this. Finally, you discuss the results – a set of three recommendations from your article and downloadable resources.

The most common problem faced when communicating research is getting bogged down in the minute details of a paper or report. This can only be expected from a person who has worked closely on a subject for a number of months. And on the whole, research, academic or otherwise, usually doesn't fit the tight, aforementioned narrative arc. There are rarely simple resolutions. And there are always caveats. Figure 6.16 offers a humorous take on the arc of an average academic paper.

Try telling your research to a friend or colleague unfamiliar with the details of your research. Ask them what interested them most and what seemed superfluous. Was there a central image of a person or subject that may form the basis of a photo essay on your blog or a one-minute YouTube video? Hone in on a specific person, scene or event that best illustrates your findings rather than trying to tackle your whole paper or report.

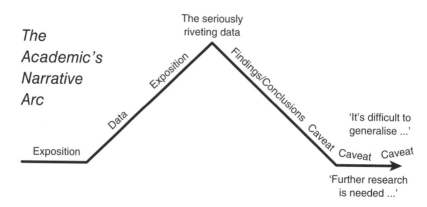

Figure 6.16 The academic's narrative arc

Credit: Source: Cheryl Brumley. It originally appeared on the LSE Impact Blog (CC BY 3.0)

Question 4: Who are your target audience and what are your aims?

If you are using photos and videos on social media as part of the dissemination and impact phases of the Research Lifecycle, you should already be considering who your audiences might be and what expectations they might have. Audiences and the actions you want your audiences to take will influence the style of the content you produce and how you use social media to promote it.

For example, if you are trying to reach teenagers in San Francisco to promote your project's website about new mental health initiatives in the city, consider a targeted campaign using short video clips on platforms with large take-up by this group, such as Instagram and Snapchat. If, for the same project, you also want to raise awareness of your findings for teaching networks, consider a one-minute explainer video that could be shared in teacher Facebook groups and on resource websites. Again, for the same project, if you are looking to show the impact of your project to funders and how it was supported by local MPs and mentioned in a House of Commons debate, a 30-second slideshow of this footage to launch your report plus your other achievements might be part of the path to extra funding. Think about who will be interested in looking at and sharing your videos, photos and research, and map these out or list them – the more specific, the better.

After considering the questions above, you should now have a clear idea of what raw photo and video content you will be working with, the story you would like to tell and the audiences you aim to reach. The next part of the planning process is to identify which types of photo and video on social media might constitute the best fit for your goals. The following section outlines a selection of different types of photography and video projects and includes real-life examples created by researchers, organisations and campaign groups. This is the time to start gathering more concrete ideas about what your photos and videos on social media could look like, before we get on to the actual creative tools stage which comes up in section 6.4.4.

6.4.2   Different types of photo

Portrait photos and selfies

- Portrait photography that captures images of your subjects, research team, participants, or others is a good (and easy) first step to consider at any stage of the Research Lifecycle. Of course, in the age of social media, we've become all too familiar with one form of portrait photography: the selfie. And indeed, if you're carrying out a participatory style project, expect a few dotted within each participant's collections.
- Portrait photography is effective in that it produces often very striking, individualised images, which capture mood and emotion. Gerry Badger in *The Genius of Photography: How Photography has Changed Our Lives* (2007: 169) sums it up: 'Of all the genres of photography, the most charismatic, and therefore the most difficult to resolve successfully, is

the portrait. A portrait photograph immediately grabs the viewer's attention and triggers profoundly personal responses – emotional, paradoxical and not always rational. The issues raised are complex, challenging, even treacherous, revolving around the self and its representation, identity and immortality.'

- As we saw earlier in the chapter, portrait images have been used with effect by social documentary photographers like Dorothea Lange, in photo-elicitation projects like Ameira Sawas's Pakistan project, and in fundraising campaigns for charities who encourage, for example, make-up free selfies, ice bucket selfie videos and donation campaign videos.
- A Yahoo Labs study by Bakhshi et al. (2015) found that images of humans (as compared to inanimate objects) – especially those smiling and making eye contact with the viewer – can help to drive social media engagement rates: they are 32 per cent more likely to attract comments. Keep this in mind if you're capturing an event or lecture: social media users want to see the human connection.
- Selfies and portraits work well across any platform, be it Flickr, Facebook or Snapchat.

Online photo galleries and collages

- Photo galleries can be embedded in blogs and collages can be created using free photo-editing tools. They are particularly useful if you have a high volume of photos, but remember you shouldn't post the full breadth of your collection. Curation is the key.
- Think about the theme of your photos and videos and what you want your audience to feel when they see and interact with your content. Some of the most powerful triggers for sharing are photos which elicit emotions like happiness and pride. Remember the Humans of New York UNHCR series?
- If you are using Instagram or another photo-driven platform, your photo feed can be embedded in your website and blog, making it a proxy photo gallery. An Instagram feed – like the Skateistan Instagram feed we mentioned earlier – is a photo gallery in itself, and allows users to browse images, read captions and add their own comments or likes (see Figure 6.17).
- If you are looking for participant engagement, consider encouraging submission via social media to your online gallery. Offer a prize or voucher to encourage more entries.

Quotes and image cards

- Quotes always work well on social media – you've probably seen a backdrop image with a quote from a book, film or motivational speech layered on top. They tap into those emotional responses we discussed above, such as happiness, anger or nostalgia. This format – where the image with the quote is created using Photoshop or an online image editing tool, then attached or uploaded as a jpeg or png image file – is used frequently by news outlets in their tweets, by a huge variety of brands on their Instagram feeds, and by universities and other groups on their Facebook pages.

Figure 6.17 Skateistan Instagram feed

Credit: www.instagram.com/p/80TJ5uKDqE

- Quotes and photos on social media are something we've used a lot on the *LSE Impact Blog*. They always have at least three times as much engagement as standard plain text and also creators can share a longer message than just 140 characters (see Figure 6.18).
- In an academic or research setting, you can overlay your photos with quotes from interviews with participants, statistics about your findings, questions you asked in the research, or any of those take-away standout messages you identified earlier.
- Brand your images appropriately, using your organisation or project logo, social media feeds, website URL and hashtag. This will help your audience recognise your content quickly when it appears in their feed. It doesn't need to be heavily branded – let your image and the quote be the focus of the output. The colour palette, filters and fonts used in the images you're creating should all be uniform if they are part of a series or campaign.

Figure 6.18 Quotations and photos on social media

Credit: Quotations on @LSEImpactBlog Twitter feed https://twitter.com/LSEImpactBlog/status/762589946133770240

- In Figure 6.19, we've created a fictional example using the free online image-editing tool, Pablo (https://pablo.buffer.com/). Sharing the key message of your research as an image in a tweet, on Facebook or Instagram – rather than just plain text – means followers are likely to take much more notice as it stands out in a social media feed.

Figure 6.19 Example of a quote image created using online image-editing tool, Pablo

FINDING COPYRIGHT-FREE IMAGES

There are many sites out there which allow you to share and re-use their photos for free as long as you give credit. You can use these as the background for your quotes and image cards on social media.

1   Search.creativecommons.org – search Creative Commons images based on key-words and you can filter by licence.
2   Pixabay – over 630,000 free public domain images. All images on Pixabay are released free of copyright under Creative Commons (CC) (Public Domain).
3   Serendip-o-matic – this website allows you to pop in text and then it generates public domain images from around the web based on keywords used in the text.
4   Flickr Commons – contains images that have been contributed by more than five dozen libraries and museums around the world and can be filtered according to crea-tive commons licence.
5   Getty Images – Getty has opened up free, non-commercial embedding of 35 million of the images in its stock photography database. You have to follow their terms of use, which don't allow reformatting or removing Getty image watermarks.

6.4.3   Different types of video

Micro video including Gifs and Vines

- Compatible with Facebook, Twitter, Pinterest and more, gifs are animated repeating images. Gifs of funny incidents or trending cultural moments are shared frequently on light news and entertainment sites and blogs, and are a part of what makes them so popular. They are used as statements, replies or comments in online conversations to convey reactions in a fun, creative and succinct way.
- In an academic research setting, gifs are probably not the most appropriate way to share videos from your project or in the promotion of research findings because of their frivo-lous associations, but it's good to be aware of their popularity in online culture.
- One area in which gifs can be used for research dissemination is in data visualisations. By creating an animation from a selection of stills – for example, showing trends on a chart – and sharing this through social media or your blog, your audience can quickly observe conclusions about growth or other trends. See Chapter 4 on data visualisation for more ideas here.
- Vine – the short-form video-sharing service where users can share six-second-long looping video clips – has been used in science and academic communication to a small extent. Joining this new video platform immediately allowed NASA to become a global leader in experimenting with social video. Adding Vine content to his already wildly popular Twitter feed @astro_reid, astronaut Reid Wiseman's space Vines of life in the International Space Station, such as harvesting space lettuce, averages over 30,000 Twitter likes and over half a million Vine loops, or plays. Joining new platforms like Vine

and Instagram certainly paid off for NASA, allowing it to experiment with content but also to reach newer, younger audience segments, and to remain visible or increase its visibility (see Figure 6.20).

- Gifs and Vines can be made using any of the various apps and websites out there, such as Makeagif.com or loopfinr.tumblr.com.

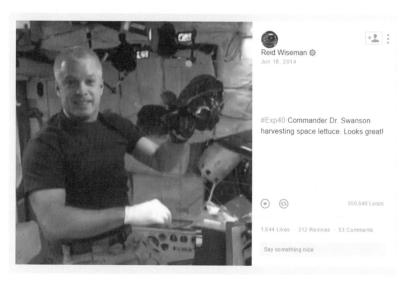

Figure 6.20 Astronaut Reid Wiseman shows off a home-grown space lettuce in the International Space Station via a Vine video clip

Credit: https://vine.co/v/MTWm7XLW2Jh

Video lectures and livestreaming

- Live video is a big trend for all the major social media platforms and will continue to generate high levels of engagements and shares. Facebook found that its users enjoyed watching live video so much that it adapted its algorithm for deciding the order in which users see content in their news feed to push all live content to the top: 'Now that more and more people are watching Live videos, we are considering Live Videos as a new content type – different from normal videos – and learning how to rank them for people in News Feed. As a first step, we are making a small update to News Feed so that Facebook Live videos are more likely to appear higher in News Feed when those videos are actually live, compared to after they are no longer live. People spend more than 3x more time watching a Facebook Live video on average compared to a video that's no longer live. This is because Facebook Live videos are more interesting in the moment than after the fact', write Vibhi Kant, Product Manager, and Jie Xu, Software Engineer, on the Facebook blog (2016).

- Video lectures are an increasingly popular format for universities and museums, as we saw earlier with examples of Nottingham Contemporary gallery and the London School of Economics both livestreaming public events as part of their public engagement strategies. They allow audiences from all over the world to be part of the event and they also extend the life of an event by giving it a place forever online and upping the project's search rankings.
- Videos can be livestreamed and/or recorded for later use. For later use they can be edited, slides can be added in, or they can be left as 'raw' footage.
- Livestreaming can be done through YouTube, Ustream, or via various other marketing providers, if you have a budget to play with. You'll also need to consider whether you have one camera or several that you can switch between.
- Livestream events can sometimes suffer from low viewer numbers if your speakers aren't particularly high-profile or you haven't promoted the livestream link sufficiently on social media. You can't just livestream and expect a 'cold' audience to be excited about what you're doing – you have to generate interest and promote the livestream in advance through carefully planned marketing.
- Launched in early 2015, Periscope allows users to stream video taken with a smartphone or tablet through its Twitter feed. While it has been generally used so far to cover events such as conferences, Periscope does allow videos to be watched again at a later date. And while it can't rival more sophisticated explainer videos and animations, in our experience, live video applications like Periscope can be powerful tools for presenting academic commentary to a wide audience in a 'talking heads' format.
- On the evening of the 2016 EU referendum results in the UK, LSE held an event which brought together academics and the research community for a discussion as the results came in. In order to bring out the debates taking place in what was a closed event, LSE's BrexitVote blog team interviewed some of LSE's academic experts using Periscope. These videos would then be embedded into the team's live blog feed. Using only a smartphone and a small tripod, Periscope allowed the team to produce ten video interviews of reasonable quality at virtually no cost. These interviews were viewed live nearly 1,000 times – not bad for a few hours while also liveblogging! One of the benefits of Periscope is that it also allows viewers to comment or 'like' the video as it is playing live; these interactions are viewable by the person taking the video, which can then be relayed verbally to the interviewee. This means that videos using formats such as Periscope can be much more interactive then traditional broadcasts. While the audience for the videos was anyone who was generally interested in the EU referendum, the marriage of Twitter and video did allow more specific audience targeting. For example, interviewee Professor Luis Garicano is also the founding editor of *NadaesGratis.es* – the most widely read economics blog in Spanish – and is the economics spokesperson for the Spanish political party, Ciudasanos. Just prior to his Periscope interview, Garicano was able to inform his more than 50,000 Twitter followers that he was about to be speaking live, and 250 of them tuned in.

Video interviews and mini-documentaries

- On the low budget end, video interviews can be conducted using Skype or Google Hangouts, or a camera and mic set up in a room, perhaps as part of your collaboration or primary research phases of the Research Lifecycle.
- At the other end of the scale, video interviews can be highly produced, scripted promotional material for social media that you might use in the dissemination and impact phases, as with the #LSEBrexitVote video interview series at LSE. Having an idea about your aims and any audience expectations should shape the format of your video interview.
- Examples can include interviewing event speakers and special guests.
- Video interviews allow you to explore topics unfamiliar to you and to gain as much from the process as your viewer, or, as you grow your content and feature new guests in a video interview series on social media, you also expand your network of researchers and academics in your field or related fields.
- Another step up on the production ladder would be a mini-documentary – perhaps three to four minutes long to take into account shorter social media attention spans – that you can share through your social media feeds and your website or blog. This will only really be a good option if you record video in the field with a documentary in mind. If you're a novice, you may want to consider hiring a professional producer or editor.
- There is always a temptation with video to want to tell the whole story. But the rule of engaging social media video content is the shorter the better. Even the experts at NASA found out the hard way, as John Yembrick, a social media manager at NASA, recalls: 'There was a test firing of a rocket engine … we did an 8-minute video and it bombed online. We pulled out a 45-second bit for social and it did great' (Yembrick, in Vazquez, 2015).

Stock footage sequencing

- This is a type of video that is used frequently by media companies for videos shared on social media. It requires the use of archival shots or video footage from photo agencies like Getty. For example, AJ+, Al Jazeera's online news and current events channel aimed at millennials, frequently constructs one-minute videos for its Facebook channel (www.facebook.com/ajplusenglish/) purely through available stock and archival footage, paired with text on screen.
- The advantage is clear: you can produce a coherent video and narrative without having to shoot a single shot. And given the sparse text needed on screen, the scripting is light and straightforward. Sources for free footage include Creative Commons videos on YouTube and the Prelinger Archives, which stock a huge collection of ephemeral films.
- You can also make use of still photographs either exclusively or interspersed with film footage. Additionally, adding movement through panning and zooming (known as the 'Ken Burns effect') will bring more life to your still image.

6.4.4 Tools for making your own digital photos and videos for social media

Now that you have a clear idea of the photos and videos for social media that you can create for your project, it is time to move on to the tools and equipment you'll need to carry it out.

Choosing a camera: Smartphone or DSLR?

The most obvious place to begin your photo or video project is with a camera. The types of skills, equipment and post-production tools vary so widely, depending on the desired quality and length of your project, that we have broken down each into an easy-to-digest spectrum chart, much like the one featured in our podcasting chapter (Figure 6.21).

Figure 6.21 Spectrum of cameras from simple to complex

Credit: Photo of Camcorder from JVC America (CC BY 2.0), Flickr.com; Photo of DSLR Camera from FotoCastor (CC BY 2.0), Flickr.com; Photo of Action camera from John Biehler (CC BY 2.0), Flickr.com

Most readers will already own a smartphone, the most simple on our spectrum in Figure 6.21. It's at the simple end of our spectrum for a reason. First, you can post videos with ease to social media outlets through one-click sharing, or utilise one of the many photo and video editing apps discussed in the following section with the literal swipe of a finger. Second, camera quality improves with every smartphone iteration, with top-of-the-range smartphones now coming with a minimum of 15MP on their main camera and 5MP on their selfie camera. The wide lens and panoramic options give you a professional finish with little investment beyond your current mobile phone plan. Third, because of their ubiquity and size, a smartphone minimally interferes between you and the subject.

However, photos and videos take up a lot of memory and a smartphone may not be suited to a multi-shoot, long-term project. You may also require a more powerful optical zoom. A compact digital camera can provide both of these amenities in addition to other features, such as optical image stabilisation (reduced blurring, although some high-end smartphone cameras and apps do have this now) and a prolonged battery life. If you are looking solely to use a video feature on a camera, consider a GoPro. It is also fairly simple to use and set up and it imbues projects with a rugged, first-person viewpoint. GoPros have a long battery life, usually up to a few hours, so they are great for long trips in the field. You can also control action cameras like GoPros remotely via WiFi or Bluetooth.

At the more complex, and indeed more expensive, end of the spectrum, we find camcorders (solely for video) and DSLR cameras. The former provide a great zoom and, unlike the GoPro, they have in-built audio recorders. They also provide manual control settings. If you learn the basics, you can achieve a more professional finish without the clunky effects of auto-focusing.

For both photos and videos, a DSLR camera like a Nikon or Canon, though pricey, is perfect for those looking for more manual control, interchangeable lenses and a sharper depth of field (meaning distinction between foreground and background). You should familiarise yourself with the camera before choosing to invest in expensive equipment – which can sometimes be a hassle. You can achieve a more professional finish on a camcorder and DSLR camera but the advantages are minimal if you stick to the auto-settings. Free courses and walkthroughs on YouTube abound as well as paid courses on online learning platforms like Lynda.com (where you can start with a free trial).

A quick word on sound: watch any first-year film student's showreel and you are almost certain to find bad sound. Sound is usually the last thing people think about when they shoot video. Many rely on the in-built speakers to do the work for them. However, achieving good sound (assuming your video requires sound) is half the task of filmmaking and in-built recorders often don't cut it. If you're aiming to conduct interviews, it is worth investing in a good microphone. The previous chapter on podcasting provides an excellent foundation in sound which is also applicable to video. Microphones can link to cameras via an XLR or phono cable or they come in mini versions that attach to DSLR cameras or camcorders much like the Rode VideoMic Pro R.

Photo-editing tools

The growth of photo- and video-driven social media platforms has sparked the creation of many free online photo-editing tools. These tools have been specifically created with optimum social media sharing sizes in mind, with ready-made templates for Facebook, Twitter and other platforms. They have easy upload, drag and drop features and give users the ability to apply filters, add text or logos, create collages and make some of the Photoshop-style fine-tuning and editing done by pros, but with far easier functionality. Every few months a new online tool pops up, but the following are some of our tried-and-tested favourites for editing photos and creating professional-looking collages on a budget or with no budget, starting with the easiest to use.

Readers will note that we haven't included programs like Photoshop in our tools section. This is because we feel they can be quite cumbersome and time-consuming; instead, we love the quick social media sharing offered by the free online tools below. We've also left out how to take a 'good photo': there are thousands of books, blogs and courses out there from which you can learn.

These three tools are a great place to start and are similar in many ways: they provide ready-made templates and allow users to customise the look and feel of their work. With these apps you can achieve a high-quality finish without a huge investment of time and money.

Pablo by Buffer (https://pablo.buffer.com/app) helps users 'create beautiful images that fit all social networks perfectly', and it doesn't get much simpler than that. Pablo allows users to upload an image or use one from its library of 50,000 copyright-free images, overlay with a quote or any text of your choosing – perhaps that key quote from a blog post, a standout percentage from your findings, or a caption for your photo – then select the size and network where it will be posted.

PicMonkey (picmonkey.com) wants to help you 'make photos that'll set everyone's "Like" buttons on fire'. A great tool for quick resizing and cropping, PicMonkey's strength actually lies in the amount of options for filters, touch-up tricks, lighting and other one-click effects. A wrinkle remover and teeth whitener (essential image-editing functions in the age of the selfie, surely?) add to the Photoshop-style features, and a red-eye remover offers that quick fix. Collages are another of the platform's strong points.

Canva (canva.com) is the most impressive of our three recommended photo- and image-editing tools, 'designed to empower you to create incredible designs – everything from social media graphics, posters and marketing materials to postcards, invitations, different document types and presentations'. It has a huge selection of templates with custom dimensions. It easily walks users through selecting a layout or theme, editing text and background images.

If you're interested in using similar tools to create infographics and data visualisations, see the corresponding section in Chapter 4.

Video-editing tools

Video-editing tools also range from basic software and apps to native computer programs like iMovie and Window Live Movie Maker, all the way to more expensive, professional software like Adobe Premiere and Final Cut Pro and Final Cut Pro X. In the box overleaf, we provide a summary of several apps that allow for quick editing for social videos. We advocate the use of these apps to make social videos before graduating to the more complex software so that you can test your level of interest in DIY film editing.

RECOMMENDED FILM-EDITING WEBSITES AND APPS

- Many readers may already be familiar with **YouTube video editor** (www.youtube.com/editor). It is not as diversified an editor as some of the others listed, but it is useful for those who want to simply get their feet wet.
- **Magisto** (www.magisto.com) claims to 'Automatically turn videos and photos into beautifully edited movies'. With this tool you can add themes and music to your videos with relative ease.
- With **PowToon** (www.powtoon.com) you can make videos without shooting a single shot. Powtoon helps you create animations and animated presentations within minutes through a drag and drop template design.
- **Pixorial** (http://lifelogger.com/pixorial) combines video editing and sharing. Users can organise their video and stills and arrange them easily in sequence.
- **VidLab** (https://museworks.co/vidlab) is a great app for creating videos to share on social media networks. It is a multi-track, multi-clip video editor that allows you to easily create videos and photo stories by adding text, artwork, music, video, sound effects, overlays and voice-overs.

Using livestreaming apps like Periscope and Facebook Live

For many researchers working alone or with limited time and financial resources, producing video of a very high quality is simply not an option. But there are alternatives which can be very effective at very low cost, especially if they are used in certain contexts. These alternatives include livestreaming services through Periscope, Facebook Live and Instagram Stories. All of these allow users to share live video with other users who have tuned in.

Each platform will provide helpful hints and tips and we don't want to repeat all of those here (mostly because video technologies are moving so fast that who knows what will come along next), but in the following box is our checklist for planning, setting up and sharing a successful livestream video on Facebook Live or any other platform.

CHECKLIST FOR LIVESTREAMING

- **Promote** your livestream in advance: tell your Facebook, Instagram or other followers on several occasions and with a lot of notice about your forthcoming livestream. Use non-social media promotion and marketing such as through newsletters or your blog or podcast.
- Just like if you were creating a video or a podcast, you should have some kind of **plan** or **script**. For example, for the live video tour of your research lab, to be broadcast live to classes of teenagers interested in studying at your university, you should prepare your colleagues, brief them on questions and think about what in your surroundings is of most visual interest.

- Most livestreaming apps will have a private or 'just for me' setting, allowing you to do a **trial run** without all of your followers seeing you fall down the stairs.
- Write an **engaging description** of what you are livestreaming in the caption or update area and use this to persuade people to keep watching: what will they be able to see? Who can they meet? Who will be interviewed?
- **Change perspectives** by switching between selfie view and the main camera. You can also flip between portrait and landscape depending on what suits the view best.
- You need to make sure your event takes place somewhere with fast broadband and a **strong Wi-Fi connection.**
- Consider buying a **tripod** or other tools, especially if you're taking followers on a tour. If your budget is spent, try putting your phone or tablet against a stack of books.
- Pay attention to **lighting and sound**. Your audience will accept that livestreaming on these apps produces content of a lower quality than something more professional, but nobody will watch something that they can't see or hear.
- The beauty of live videostreaming through these apps is the social element: those watching can **ask questions** and make comments – to which you can respond on camera. As with any social media comment function, you may want to consider moderation and guidelines.

Hiring professionals to create your photos and videos

Photos and videos have whole businesses built around their creation and you should accept that what you might be able to create using the tools we've discussed above may not be as slick as something made by a professional. We would recommend using professionals to create your content if you are looking for a highly polished finish, if the story you want to tell is complex and you have little to no experience of creating this content – especially when it comes to video.

If visual creativity doesn't come easily to you or your skill level isn't very high, it's probably best to hire the skills of an experienced photographer or videographer, or at the very least talk to some creative professionals who can show you examples. If you are an academic attached to a university or a researcher at a large organisation, you may have access to the skills of a professional marketing team, including video producers and photographers who can work to your brief to create this content for your project. The brief should cover information about the audience you are trying to reach, the messages you want your content to convey and the platforms through which the content will be shared. This is all important information for the professionals, who will have a good understanding of the styles and approach more suitable for particular audiences. You will need to go into the meeting with clear ideas and examples of photo or video styles that you like or particularly don't like.

A real plus of working with professional photographers and video producers at your university or institution is that they will make sure that the work you are commissioning fits in with the wider branding of the organisation. You'll benefit from maximising the power of your institution's brand by having its logo featured on your work, for example, and from it being shared with a wide institutional social media following, including alumni and students. Plus, the professionals will have a clear idea of best practice, legal requirements and copyright issues, and can advise you of these.

The use of professional skills and guidance comes with a price tag, however, so explore calling in favours or ask any contacts for some pro bono advice, or make sure you portion off some of your budget or bid application to account for covering these activities.

6.5 SHARING YOUR PHOTOS AND VIDEOS ON SOCIAL MEDIA

You've found the perfect research story to tell through photos or video on social media. You've planned your project down to the last second of footage. You've spent time getting colleagues and participants on board, learning about setting up great shots, or pulling in favours from creative professionals. Now it's time to share your work on social media in a way that will increase the likelihood of audience engagement, drive awareness of your wider project and perhaps even make a bigger impact! Having a social media following in some form is a start, but there are lots of other things you should consider, which we'll now explore.

6.5.1 Sharing photos and videos through social media channels

Throughout this chapter we have given examples of how photo and video projects have been shared on specific social media platforms and services, such as Pinterest and Facebook – and we'll look at those in a bit more detail now. However, this is not to say that these are the only ways to share your photos and videos on social media. Finding the right platform to share your content will depend largely on the format you choose, the content you are able to pull together, and your target audience. We would advise picking one or two platforms to share your work through and doing this well, rather than trying to spread yourself too thinly across six or seven different platforms and doing it all badly. Instead, find out where your audiences are and go there.

As with data visualisations and infographics, **Twitter** and **Facebook** are obvious choices as prime social media channels to share your photos and videos. Audiences here are generally looking for news and entertainment, and are interested in consuming and sharing content that reaffirms their opinions and beliefs. Sharing the title to your video or photo series, or a take-away quote from the script, along with the image or video file of your work, will help to attract more viewers. Some examples of effective tweets or Facebook updates for promoting photos and videos include:

- New video from our project on political engagement shows what young people think of #DonaldTrump – watch here + VIDEO
- The reality of gender #paygaps around the world compared in this photo series and blog post + PHOTO
- Video #infographic: steady growth generates higher levels of wellbeing among citizens than boom and bust cycles + GIF VIDEO

Both of these platforms enable users to upload autoplaying video files, attracting high levels of interaction. You can paste in the YouTube or Vimeo link to your video, but best practice is to upload the mp4 directly. However, the videos will play muted – there is no sound to attract the viewer – so best practice is to add subtitles or additional floating text to any video that you upload. Subtitles can be uploaded on Facebook through a specific.srt file, or you can add them yourself at the video-editing stage. Fill out all descriptions and metadata options when uploading the files – select a clear and attractive thumbnail image, add a title, description, hashtags, etc. In the description, use call to action words such as 'watch', 'now', 'like', 'tell us' and 'you' to make your update have the best chance of engaging. Twitter and Instagram currently have a limit on the length of video you can upload to autoplay – around 60 seconds – so you could create a short trailer for any longer material.

Facebook is best for sharing...
Livestreaming
Photos and photo galleries
Short videos uploaded to autoplay
Content to be shared in specific groups and communities

Twitter is best for sharing...
Quotes
Micro videos
Trailers or clips
Content to be shared widely

YouTube and **Vimeo** are dedicated video platforms that can make for attractive portfolio bases for your video work. Create playlists of your work under different themes: this will help users navigate your work and encourage them to watch more. Just as with Facebook, fill out all the description and caption elements: these will help boost the searchability of your work. Tracking your non-academic citations with Altmetric Explorer or similar? Use the video description to link to your DOI too.

Instagram is the social media platform that enables users to take pictures and videos, apply filters and effects, and then share publicly or privately with friends and followers. For a platform favoured by over-exposed Kardashians, Instagram might not seem like a natural fit for disseminating anything related to academia or research, but it has been quickly embraced by organisations including The National Geographic, UNICEF USA and the Mars Curiosity Rover, who are able to showcase their educational findings, campaigns and news in a very engaging way (Paton, 2013). Two institutions heavily involved in knowledge creation – libraries and universities – have also taken up Instagram in large numbers. Libraries such as New York Public Library (@nypl) and The British Library (@britishlibrary) use Instagram to show off their collections and engage users in discussions about their favourite materials (Mollett and McDonnell, 2014), and universities such as LSE (@londonschoolofeconomics) and NYU (@nyuniversity) are using the platform to celebrate achievements and share inspiring images of cutting-edge research (Mollett and Fazal, 2014).

Instagram is owned by Facebook and the two platforms connect easily, so a quick way to grow your Instagram following is to re-publish your Instagram images in your Facebook groups or pages. Take advantage of the long description on each Instagram photo to add a detailed description of your content: examples include adding a description of the topic of your research, asking a question, adding a call to action (click here, watch here, etc.), or adding a URL that you want your followers to visit. As of the summer of 2016, Instagram's captions did not allow for hyperlinked URLs, though users do often use their biography space to add a link to their latest blog article, writing 'see link in bio' in their photos caption.

As with Pinterest and Twitter, a big part of using this platform is the social aspect. Whatever area of research you are working in, follow others with similar interests and double tap to like their photos.

Instagram is good for sharing...
Capturing instant photos and videos
Content aimed at young people
Micro videos
Light-hearted content

Just like Twitter, use specific and relevant hashtags in your photo and video captions when you publish on Instagram so that your content can be found more easily by users. Also use specific, unique hashtags for individual events or longer campaigns for your project, so that you can easily showcase all your content together.

Pinterest, the visual bookmarking social media platform, used to have a reputation for being a niche social network filled only with photos of kale smoothies, collages of cute kittens and photo-guides to IKEA hacks, but it's come a long way since its launch in 2010. Pinterest now has over 100 million users and has become hugely popular for businesses and content publishers; for many brands, it is the second biggest social driver of web traffic after Facebook. In terms of its user demographics, 42 per cent of online women are Pinterest users, compared with just 13 per cent of online men; 34 per cent of adults 18–29 years old and 28 per cent of adults 30–49 years old use Pinterest (Pew Research Centre, 2015). Each Pinterest user can share content by re-pinning images from other pinners that they follow, uploading their own images, or creating a pin from an image or video on an external website. As well as sharing your own project photos for Pinterest, you can also embed yourself within the Pinterest research community by re-pinning pins from other organisations, campaigns, libraries or universities. This will help to make your channel a valuable resource and build authority and positive associations with your project.

Pinterest is best for sharing...
Photos with quotes
Infographics or collages
Videos
Community images related to your work or theme

Pinners who click to look at a pin in more detail can then click again through to the source of the image, which can be a blog post, a news article or a video. When naming and describing your images and Pinterest boards, keep SEO in mind and front them with keywords. Being keyword rich and very specific about what you're pinning means they can be found easily when users are searching for pins to pin or pinboards to follow. Many pinners use Pinterest as a visual search engine, which is another reason why those keywords and short snappy names are important.

Make sure you add a Pinterest social sharing button to your website that will sit alongside 'Like' and 'Tweet' buttons, and in your options make sure you can add these under most images if you're happy with that.

TIPS FROM LESTER ANDRIST, *THE SOCIOLOGICAL CINEMA*

First, popular pins tend be ones that convey information that is timely and relevant. When the Black Lives Matter movement makes news, it's not surprising that pins about Black Lives Matter do fairly well.

(Continued)

(Continued)

Second, pins that do well tend to reveal something usually hidden or not often discussed. One pin that has done well features a homeless man trying to sleep beneath a statue of Ronald McDonald, and it appears as if the clown's foot is stepping on the man's head. I think this image strikes people as revealing something about economic inequality and consumerism.

Third, successful pins can get away with a little academic jargon but not a lot.

Hashtags across all of the platforms we've discussed are an effective way to reach beyond your individual network. Be sure the hashtag is active enough to reach people but not so populated that your content will get lost in the noise. Consider using both general hashtags and hashtags specific to the topic of your work – for example, general hashtags that browsing users might search for would be #video and #photography, and more specific tags that users undertaking a more targeted search might use could be #history #womenshistory #genderpaygap and #feministinterviews, for example.

6.5.2 Sharing photos and videos through blogs

As with the infographics and data visualisations, having an accompanying blog post for your photos and videos can provide a useful anchor for your multimedia content. An accompanying blog post that gives a short description of your photos or videos can add to the search engine optimisation of your work. A blog post is a social media-friendly method of dissemination – with blog platforms built for one-click sharing and easy social share counting and, to add SEO weight, with keywords most likely scattered throughout your accompanying write-up. When it comes to sharing a blog post about your photos or videos through social media, you'll ultimately be sharing the URL to the blog post, which will give readers a base to come back to and read more about your work and your project.

Most blogs are set up to easily embed YouTube and Vimeo links, so it may be best to upload your mp4 to these platforms first. As mentioned in Chapter 3, sharing images is always encouraged on blogging platforms, but if your photography project is a collection of images, consider uploading to Flickr and embedding the Flickr link to the blog to help direct to the full collection. Image- and video-sharing platforms like YouTube and Flickr have their own established community of users which may be worth engaging in further for remotely-connected audiences.

For an in-depth look at all things blogging – including choosing where to blog and how to blog – see Chapter 3.

6.5.3   Sharing photos and videos in the 'real world'

Don't forget that the 'real world' exists too! After you've scripted, recorded, edited and published your photo or video content on social media, be prepared to discuss what you've created and how you did it at every opportunity. Conference presentations, public lectures and festivals are an obvious venue to tell an engaging story about how you used these media formats at any stage of the Research Lifecycle, but there are plenty of other spaces. Many universities organise informal Show-and-Tells, TEDx talks and Fringe-style events aimed at public audiences. Consider where your findings might provoke discussion and debate and look to see if there are any opportunities for engaging these audiences in dialogue.

Promoting your photo and video content while networking in person can be of great help to your project. You could include the YouTube, Instagram or Twitter profile on your business cards or presentations.

6.5.4   Sharing photos and videos through targeted emails

To make the most of this channel for sharing your photo and video content, use a newsletter tool such as MailChimp or Campaign Monitor to create a design where your visual content is the first thing your readers see when they open the email. Videos can be embedded so readers can watch straight away whereas gifs will automatically play. Use the email subject line to add a call to action and make opening the email an irresistible proposition.

As with a data visualisation, add a paragraph to briefly highlight some of the key messages and include links back to your project blog post or website, social media channels and any full reports. You could use standard GMail or Outlook email software but using a more professional tool will allow you to track the number of opens, clicks and other engagements within your emails – useful information to know for future mailouts or project reports. Overall, this is a fairly simple yet effective way of getting your content seen, and will complement your social media strategy nicely.

When it comes to building your email lists, add and segment colleagues past and present, research contacts you've met through networking and journalist contacts if you're seeking media coverage. Segmenting the lists means you can send separate emails and invitations to these groups and tweak the messaging and tone accordingly. A quick note on building subscriber lists: do not spam 'cold' contacts that you've never met or interacted with, but ensure that your contacts would instead be happy to hear from you. And don't forget to share on relevant mailing listservs.

Whatever method you pursue, make sure to coordinate a plan for sharing your photo and video content. Focus on what specific audiences will be most interested in and will benefit the most from your content and look to build attention from there. In almost every case, an audience does not happen overnight. Continue to pursue opportunities where conversations, both online and offline, can occur and make sure to make notes on what works for the future. And if you are looking for further guidance, Chapter 7 includes more detail on choosing the right platform and planning a coherent social media strategy.

6.6   CONSIDERATIONS

As is the case with previous chapters, copyright laws vary widely from country to country. It is beyond the scope of this book to offer legal advice, although below are a few things you should think about before using another person's work in your video project and releasing your own original content using an open license:

- Note: laws are less lenient around the use of someone else's photography, so we encourage you to use Creative Commons photographs only, unless you have access to a subscription-based image bank.
- When it comes to using portions of video and video-based performances, be aware of the fair use (or fair dealing) laws in your country. As stated previously, fair use/fair dealing limits the exclusive right of copyright holders, allowing for other content makers to play or feature a portion of someone's work for their project. There are still restrictions on length for videos, purpose and character of the use. As pointed out in the previous chapter on podcasting, the issue of allowed length in the case of video or audio is often hard to interpret so it's best to get legal advice specific to your case. Generally, these laws allow for short, relevant segments of a collective work to be used for education purposes, criticism or review.
- For more information on fair use in the US, visit the US Copyright Office's website at: www.copyright.gov/title17/92chap1.html#107. In the UK, visit the UK Copyright Service website at: www.copyrightservice.co.uk/copyright/p01_uk_copyright_law. And in Australia go to the Copyright Council website at: www.copyright.org.au.
- Similarly, when you share your own original images, consider using an open attribution licence like Creative Commons (denoted by the acronym *CC*. You may also see *BY* which stands for the need for the correct attribution of your work) to encourage wider sharing. By blocking the use of images by commercial outlets such as newspapers, websites, social media, YouTube and books, researchers block the most effective ways of disseminating their images and research. Researchers may wish to publish messages asking journalists to contact them for permission, but UCL guidance (UCL, 2016a) notes that picture editors working to tight deadlines 'will just move on to something else, losing you an opportunity to increase the impact of your research'.

- Every institution and project will differ in the support they give to Attribution and Commercial licences, but we draw inspiration from NASA, whose ubiquitous images are by law released to be re-used, adapted and remixed by the public. Ray Villard, head of news at NASA's Space Telescope Science Institute, notes that those who try to restrict the re-use of images 'are antithetical to public communication and engagement in this social media age. It is guaranteed to kill any publicity for their research. Reporters and bloggers are on deadlines and will not waste time to address copyright issues. Reporters will simply find free sources – like NASA.'

6.7 FURTHER READING

Blyth, M. (2012) 'Five minutes with Mark Blyth: "Turn it into things people can understand, let go of the academese, and people will engage".' [Blog post]. *LSE Impact Blog*. March 9. Available at: http://blogs.lse.ac.uk/impactofsocialsciences/2012/03/09/five-minutes-with-mark-blyth [Accessed 14 November 2016].
Here Mark Blyth, Professor of Political Economy, explains why being unreadable helps economists get their message across, how fan and hate mail have become part of his professional life and how his latest project illustrates that there is a market for academic ideas.

Mollett, A. (2013) 'Using Pinterest to create reading lists: a step by step guide.' [Blog post]. *LSE Impact Blog*. September 27. Available at: http://blogs.lse.ac.uk/impactofsocialsciences/2013/09/27/using-pinterest-to-create-reading-lists-a-step-by-step-guide [Accessed 14 November 2016].

Prosser, J. (2005) *Image-based Research: A Sourcebook for Qualitative Researchers*. London: Routledge.
A self-styled 'sourcebook for researchers in the field' conducting image-based research.

Tinkler, P. (2013) *Using Photographs in Social and Historical Research*. London: Sage.
A useful book that outlines the conceptual and theoretical that underpin photographic research.

Vazquez, L. (2015) 'How NASA turned astronauts into social media superstars.' [Blog post]. *Popular Science*. October 9. Available at: www.popsci.com/how-nasa-trains-astronauts-for-instagram-and-beyond [Accessed 14 November 2016].
The story behind NASA's success on Instagram and other social media platforms.

Vimeo Video School. Vimeo. https://vimeo.com/blog/category/video-school
Engaging and comprehensive vlogs on how to shoot and cut your own videos.

7

DIGITAL STRATEGIES FOR RESEARCH DISSEMINATION, ENGAGEMENT AND IMPACT

Not everything that can be counted counts, and not everything that counts can be counted.

– Cameron (1963: 13)

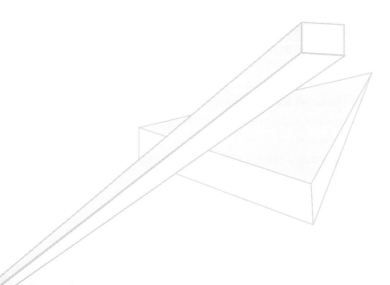

The previous chapters provide a solid basis for any researcher to approach, create and share a wide variety of research content. But the specifics of individual projects and research settings will shape the ways in which these tools and methods are actually applied. The practically limitless possibilities for the content and the sharing may seem overwhelming – and it is fair to say we have all been overwhelmed by the vastness that is social media at some point. To bring this book to a close we take a look at how individuals and research teams can more systematically and strategically approach their digital content and online identities.

Chapter 1 looked at the influential and expansive role of social media in our lives. But rather than viewing social media as an appendage to research, Chapter 2 has shown how social media can be integrated across the Research Lifecycle, which in turn can help research aims rediscover the fundamentally social process of research. Chapters 3–6 demonstrated how to actually go about creating engaging content. Through this lens, we argue social media can be viewed not as another burden to the lives of researchers, but as a relatively low-cost, high-value opportunity to engage and build relationships. But we also recognise the limitations that researchers face. At a time when budgets are stretched worldwide and UK public investment in research in particular continues to flag in real terms, organisations and projects need to make the most out of limited time and resources (Rohn et al., 2015). Communicating research exists within this imperfect structure. Adding to this difficulty is the fact that research communication activities often go unrewarded or are even derided by more senior colleagues (Gruzd et al., 2011). Incentive structures in academia for digital activity, for example, have yet to fully catch up in a way that encourages these efforts (Meyers, 2014).

Making social media tools work for your research aims requires deliberate action and deliberate thought throughout the project. Social media are not something to be left to the end of a project. Researchers and institutions can foster intentional strategies to boost their online presence to maximise successful engagement. This chapter will outline some potential strategies by first looking at how to approach existing platforms. Then it will consider the many ways of measuring various social activities. Finally, it will end with a brief reflection on how to navigate the pitfalls and risks of social media.

7.1 CRAFTING A COHERENT SOCIAL MEDIA STRATEGY

Given the stretched nature of researcher workloads and research budgets, getting the most out of your social and digital media activity is incredibly important. So how can individuals and research teams understand their own communication efforts and act strategically? In order to answer this question, we return in particular to the final three aspects of the Research Lifecycle: Dissemination, Engagement and Impact. Depending on the specific aims of any given project, a closer interrogation of these three activities can

	Facebook	Twitter	LinkedIn	Pinterest	Instagram	YouTube
Number of monthly active users	1.65 billion	310 million	100 million	100 million	400 million	1 billion
Key audiences	87% of 18–29-year-olds worldwide use Facebook. 77% of adult women and 66% of adult men use Facebook	37% of 18–29-year-olds use Twitter. 24% of adult men and 21% of adult women use Twitter	31% of 30–49-year-olds use LinkedIn. 28% of online women and 27% of online men use LinkedIn	34% of 18–29-year-olds use Pinterest. 42% of online women and 13% of online men use Pinterest	53% of 18–29-year-olds use Instagram. 29% of online women and 22% of online men use Instagram	Almost a third of all people on the internet are YouTube users
Used for	Connecting personally with friends, family, and topics that we care about	Finding news and entertainment, online networking, and feeling connected to large trending debates	Professional networking and job hunting: an online CV	Curating images connected to hobbies, ideas and aspirations	Popular network for instantly sharing photos and videos. Enormous user base amongst young people around the world	Dedicated video platform where any individual, brand or organisation can show off their content in a shareable and attractive format
Why it is good for research	Huge user base. Great for building public and private groups around themed topics. Also great for sharing video	Very popular social media network for connecting with other researchers and also journalists. Follow hashtags around research interests	Good for sharing blog posts, developing a professional network	For anyone working with infographics and images, this is a useful platform for connecting with students, colleagues, and smaller interest groups	Offers a light and accessible way of sharing photos and videos with hard to reach younger audiences. At the forefront of social media video and photo sharing trends	Playlists show users more of your video content, and a strong search and tagging function means your videos will be recommended to others

Figure 7.1 Definitive guide to social media platforms for communicating your research – infographic

Source: https://blog.hootsuite.com/social-media-statistics-for-social-media-managers

help researchers orient their associated digital strategies, monitor their development and better understand their value and influence throughout the process. To be sure, an effective communications strategy is useful throughout the Research Lifecycle and should be considered long before you are looking to disseminate your work. But especially for teams looking for wide exposure and social influence, a systematic approach to choosing platforms, setting goals, understanding your audience and actively listening will help ensure you are laying sufficient groundwork for your social media activities.

When you are considering what social media platform to invest your time in, you should first think about the architecture of the social media platforms themselves. Again, we are not advocating for specific platforms, but rather highlighting ways to understand how social media platforms operate. In Chapter 1 we looked at how audience size and shareability are two primary ways to categorise various tools, content and networks. Based on the social sharing sections in Chapters 3–6, Figure 7.1 is a quick overview of the most popular social media platforms we have discussed and why they are useful for research purposes.

But this infographic is not intended to limit experimentation. Rather, it is intended to help researchers situate their own activities and to continue to explore new and emerging ways to engage online.

7.1.1   What are you looking to get out of online activities?

With organisations under increasing pressure to demonstrate their relevance and contribution to society, a deliberate digital strategy is vital. Institutions and central communications teams may provide some help and support in pointing your digital efforts in the right direction. But a digital strategy is not necessarily just enacted from the top down. With more and more individuals using digital tools for everyday research and educational practice, digital strategies are continuously enacted, whether this action is conscious or not. A deliberate consideration of what you are looking to get out of your digital activity, and how your practices fit within this, can offer a number of benefits. Creating a coherent strategy that aligns both the institutional and individual motivations for social media use is a worthwhile activity.

Mark Reed (2016) has written on the importance of setting goals for social media use in his *Research Impact Handbook*. He writes, 'Researchers are in a unique position on social media, because we have easily verifiable credibility as authoritative voices' (Reed, 2016: 162). This credibility can be seen as a distinct advantage, especially on social media where sources are often dubious. But without a clear plan, Reed argues, this advantage can be lost. He suggests seven questions (paraphrased below) that can and should be asked and answered by researchers at the start of their social media projects:

1. What do you want to achieve?
2. **Who are you trying to reach?**
3. What content are you working with and how can you repurpose this?
4. Who can you work with to make your social media use more efficient and effective?
5. **How can you make your content actionable, shareable and rewarding?**
6. How will you monitor and evaluate your plan?
7. How does your social media plan contribute to your wider research impact plan?

(Adapted from Reed, 2016, emphasis added)

By answering these seven questions, a deliberate strategy begins to emerge. In our experience, the biggest questions to focus on, and the ones we see most often ignored, are the questions on identifying who your audience is (question 2), and figuring out exactly what it is you'd like your audience to do with the content and information you provide (question 5). By narrowing in on these questions in particular, the types of social media platforms and the types of content most suitable to these actions should become more apparent.

For example, if you are conducting research on digital privacy rights, and your research findings offer insights that would be particularly relevant for parents of children aged 8–13, writing a blog post on your personal blog and tweeting it to your largely academic Twitter followers might be selling your work short. Why not identify four or five specific actions for parents and children, put together a basic infographic and pitch this research summary to a few parenting podcasts, online discussion forums and blogs?

Or, say you are researching the financial literacy and savings decisions of millennials. Your data could provide a comprehensive account of dwindling savings rates among 18–35-year-olds and the negative effects this will have on economic security further down the line. But in order to reach the audience that could benefit the most from these insights, you might consider the use of a platform like Instagram for sharing your research and provoking further thought and discussion. A photo portrait series focusing on the lives and daily financial decisions of these young people, or a podcast that lets research subjects share their own stories alongside the presentation of your analysis, could resonate more strongly than presenting the data alone.

These are overly simplistic examples and we know it is not always easy to identify who the relevant audiences are for your work. We spend more time on how to map and understand different audiences in the next section. And, of course, a strategy is just one piece of the puzzle. The content itself has to be engaging. This can be trickier, but the previous chapters have discussed in more detail how this can be done, what engaging content looks like in practice and where to go for more ideas.

Another helpful way to conceptualise your social media strategy comes from Holly M. Bik and Miriam C. Goldstein in their article 'An Introduction to Social Media for Scientists' (2013). They provide a flowchart decision tree, reproduced in Figure 7.2,

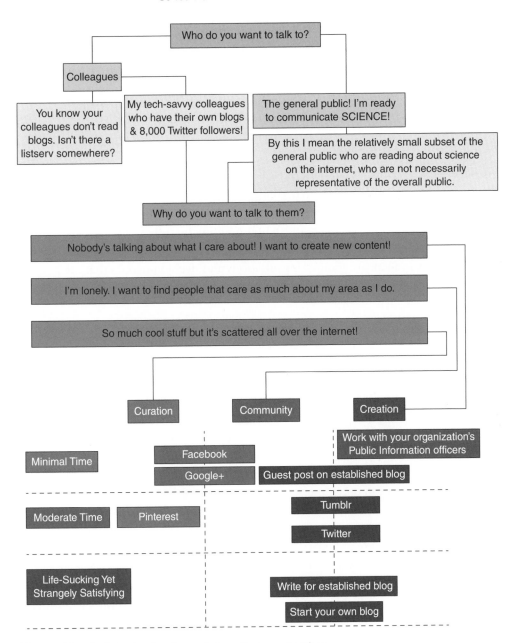

Figure 7.2 An introduction to social media for scientists

Credit: Bik, Goldstein (2013). CC BY 4.0. An earlier version of this flowchart appeared in a guest post by MCG in Nature's Soapbox Science blog (http://goo.gl/AeKjJ)

for scientists getting started with communicating research. The chart centres on two questions: one about target audience and one about primary motivations for the communication. From there, a number of strategies are suggested, depending on familiarity of online platforms and time commitment. A clear articulation of purpose for what you are looking to get out of social media use is central to both Mark Reed's and Bik and Goldstein's approaches and is a fundamental aspect that merits sustained consideration.

7.1.2 Audience mapping

We have spent a significant amount of each chapter of this book referring to the importance of audiences for your research communication activities. The medium of your research messages must be chosen with careful attention, as the media themselves will shape the content, form, reception and action resulting from those messages (McLuhan, 1994). But what exactly do we mean by external audiences? We define audience as the many different individuals, groups, organisations and decision-makers that are affected by or may be able to use your research in some way – whether for further research or the application of research.

Some may argue that the word 'audience' is itself too enmeshed in an 'old media' view of individuals as passive receivers to be applicable to 21st-century communication and research (see section 2.1 in Chapter 2 on models of research communication). There is also a body of public relations scholarship that differentiates 'publics' and stakeholders (Tench and Yeomans, 2009). Grunig and Hunt's (1984) Situational Theory, for example, distinguishes between 'publics' by identifying their awareness of the problem and the extent to which they do something about the problem. But regardless of whether you refer to external groups as audiences, stakeholders or communities of practice, we are referring here to the groups and individuals you are looking to communicate with and influence in some way in order to reach your project goals.

For some research projects, it will be immediately obvious who the primary beneficiaries and target groups are for your research and how best to go about reaching them. But for many more researchers and communicators, the question of audience will be far more difficult to answer. Especially for researchers looking to reach non-academic audiences, the concept of the 'general public' can be difficult to unpack. But this is an essential step, as an undefined audience will only lead to an undefined message and your work will struggle to connect with those who would benefit most from it. To avoid wasting time, energy and resources, steps can be taken to ensure your research content and communication efforts are not in vain. Here we look at some examples from public and civic organisations on how best to refine and review the relevant audiences for your digital projects.

Audiences are the people, groups and organisations that are affected by your research project. According to Dr Andy Williamson, founder of FutureDigital, having a clear understanding of who these groups are, who they are influenced by and who they in turn influence can help you understand the change your work makes and where best to direct your attention (Williamson, 2003: 1). He proposes concentric circles for under-standing your key external relationships (Figure 7.3). The innermost circle is the 'Direct' sphere – those you are currently working with. Moving out from there is your Indirect audience – groups you do not interact with, but who are big players in the sector or are directly affected by your work. The next circle comprises the Remote groups – those who could potentially be affected by your work but are a step removed. Finally, the Societal circle includes broad categories of people and more macro-level influences (e.g. legislation) that shape the work you do. Williamson argues that by plotting these groups and the con-nections that exist between them, you will have a better understanding of the information they are looking for and how best to reach them (2003: 11).

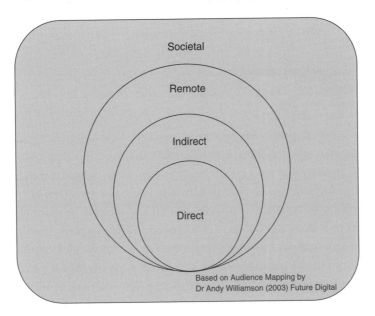

Figure 7.3 Audience mapping exercise

Credit: Dr Andy Williamson. Original © 2003 Wairua Consulting Limited & revised © 2013 Democratise/Future Digital

From this audience mapping exercise, you should also be well placed to pull together a power–interest matrix first developed by environmental researcher A.L. Mendelow (1981) to identify where your groups sit according to power or influence and interest (see Figure 7.4).

The purpose of the power–interest matrix is 'to keep track of the changing criteria by which their stakeholders judge their effectiveness' (Mendelow, 1981). With ever-shifting public priorities, research agendas and societal needs, a matrix like this can enable your team to identify what groups are interested in your findings and which actors would be able to carry your messages further. In terms of social media engagement, audience mapping and the power–interest matrix approach can also be useful for identifying what social media accounts to build relationships with and directly target.

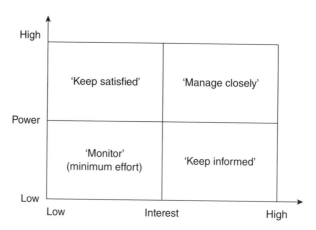

Figure 7.4 Power–interest matrix

Credit: Wikimedia, Creative Commons CC0 1.0. Universal Public Domain Dedication. https://commons.wikimedia.org/wiki/File:Power-interest_matrix.png

Stakeholder analysis approaches have been used across business, charity and public sectors for decades (Goodpaster, 1991; Ramirez, 1999; Brugha and Varvasovszky, 2000). But Reed et al. (2009) note that stakeholder analysis may mean different things for different people and, especially in the research community, there is a great deal of confusion over who is and isn't a stakeholder. One distinction that may need to be made before you start determining your audience is the level of response and engagement you are looking for. A general public awareness of your project can be achieved through a combined approach of mass and social media platforms to get you broad exposure, but if you are looking to change policy or behaviour in some substantial way, you will need to spend considerably more time understanding the motivations of the ecosystem of stakeholders, actively listening to the feedback and incorporating perspectives from these stakeholders into your research communication activity. We look in greater detail in the following sections at the emerging issues and categories of social media platforms and then consider the practice of online listening as a deliberate strategy to get the most out of your engagement with audiences and stakeholders on social media platforms.

7.1.3 Trends in social media platforms and dealing with 'context collapse'

Though we have mentioned a few of the most popular platforms for communicating research, this book has not covered in extensive detail information on specific social media platforms.

We have done this partly because a book on specific social media platforms would date incredibly quickly. But also, we are not advocating for social media tools per se; instead, our underlying argument is that through a wider embrace of social and digital media, researchers can more efficiently and effectively communicate with an interested community. Platforms will come and go, and innovative uses of specific platforms for communicating research depend on a variety of external factors, such as a target audience (who you are trying to reach), digital familiarity (how you are going to reach them) and the sensitivity and nature of the findings (what it is you are trying to communicate).

But within social networking platforms in particular, there are a wide range of features and peculiarities. Below are a few additional aspects to review which may also influence how you adopt, engage and interact with these platforms.

Public vs private

Do you want to share content with a wide unknown public? Or are you interested in maintaining a level of control over what is shared and how it is received? Some social media platforms, like Twitter, are primarily geared towards public audiences where default settings make content live for everyone to read and messages are retweeted and shared to wider networks beyond one's own following; whereas Whatsapp (mobile messaging app), Slack channels (digital chatrooms primarily aimed at facilitating team communication) and private Facebook or LinkedIn groups push towards known, specified audiences and parameters.

But simple categorisation of platforms as either public or private is highly questionable. Dave Beer (2009) has argued that a key feature of Web 2.0 platforms is their emphasis on 'participatory' (read: data-generating) web cultures. Spaces that may appear private are still very much visible to the owners of the platform and can thus be exploited. danah boyd (2007) has also written that in many areas of social life, but especially in relation to social media sites, 'What it means to be public or private is quickly changing before our eyes and we lack the language, social norms, and structures to handle it.'

Navigating this space requires deft attention not only to help us align our communications activities according to individual preferences, but also to ensure we, as participants in a complicated web culture, understand the broader societal context in which we are working. Before you post online, consider what these 'networked public' dynamics might mean for you and for your work. Bonnie Stewart (2016) urges us to ask: What do your messages convey about your institutional role and your personal identity? Are you comfortable with the blurring that occurs online between personal and professional? In what ways are some people more vulnerable than others to heightened visibility online? More on navigating these questions will be discussed at the end of this chapter.

Open vs closed

Considerations of whether a platform, and its underlying infrastructure, is open or closed have received growing attention from the scholarly community. It is more commonly recognised that websites like Amazon and Facebook are generating enormous revenues from the monetisation of their users' data. But sites like Academia.edu and ResearchGate, social networking sites where researchers can upload and share versions of their scholarly articles, have also come under mounting criticism for the lack of transparency around their business models and ethics (see Fitzpatrick, 2015; Hall, 2015; Moore et al., 2016). Many people are asking why we are choosing to allow these companies and their investors to profit from scholars' activity online when that money could be better spent on more open platforms, owned and managed by the scholarly community itself.

There is an obviously big reason why people use these closed platforms (even when it may go against their long-term interests), and that is because there are few alternatives out there that are as user-friendly and/or ubiquitously used by colleagues and the general public. But that doesn't mean that other more suitable alternatives might not emerge over time. As you consider the adoption of particular platforms, taking a look at the longer-term effects of the platform may help. Bilder, Lin and Neylon (2015) propose four key aspects to consider when weighing up the suitability of infrastructures for scholarly communication. The quote below is from Bilder, Lin and Neylon (2015), as paraphrased by Moore, Gray and Lammerhirt (2016):

1. **Governance**: They should be stakeholder governed, transparent and non-discriminatory, understanding they do not have a right to exist beyond the service provided to the community.
2. **Sustainability**: Infrastructures should be made to generate surplus, have contingency plans and, crucially, base their revenue on services not data.
3. **Insurance**: The research community needs to know it is in control and can maintain the control in the event of anything unexpected. The authors recommend that services be based on opensource software, open and reusable data and patent non-assertion. This will allow the community to inherit aspects of the service that for whatever reason can no longer be maintained.
4. **Implementation**: The trickiest aspect of these principles is how they are implemented and what kind of organisations are most suited to them. The authors assert that they naturally lead to a 'board-governed, not-for-profit membership organisation' but other models should be explored too, including centralised vs federated systems.

It would be unrealistic only to use platforms that work in the interests of the research community. But these principles can help identify what control the research community is ceding by using third-party platforms and whether it is worth investing time and money in more sustainable platforms.

Permanent vs ephemeral

A final feature worth considering in relation to your adoption of specific platforms is the permanence and stability of the content featured therein. In recent years there has been a growing number of social technologies favouring a more transitory approach where content vanishes quickly after it appears, and may even be completely anonymous, taking the notion of real-time to unprecedented levels. Platforms like Snapchat (the image messaging and mobile app discussed in Chapters 1 and 6) and YikYak (an anonymous bulletin board messaging app based on GPS location) both integrate this 'perpetual present' ethic into their core design as visual and text-based content appears and quickly vanishes after it is posted (Polonski, 2014). This is a marked change from the social networking profile page of the early 2000s as an archive-able record of fixed identity. These ephemeral platforms offer new ways for individuals and organisations to present themselves as fluid, dynamic and responsive. But how suitable is this for research where persistent content is in many ways a prerequisite for future impact?

While transitory content may feel anathema for research purposes, particularly for researchers ingrained to cite and use verifiable sources, these emerging technologies have begun to take off in Higher Education circles. In a 2016 *Times Higher Education* article, 'Is Snapchat the new Twitter on campus?', Alistair Beech, senior digital communications officer at the University of Central Lancashire, looks at the popularity of these platforms among students and how these apps are being used for student recruitment and marketing:

> [Snapchat and Yik Yak] can be essential for students adjusting to a new community and environment. ... Both networks are mobile, real-time and change according to location – providing great opportunities for universities to connect with prospective students and students wherever they are – campus or not. (Beech, as quoted in Grove, 2016)

So why are social media users turning to these kinds of ephemeral digital media? An obvious answer would be the established popularity of mobile-first platforms. Another explanation suggests it may be down to the lack of control we feel over the very dichotomies we have been discussing. The blurred identity boundaries that exist between personal and professional, public and private, and the many ways this complexity is presented and received across networked spaces, otherwise referred to as 'context collapse', has led to a further experimentation of tools that subvert traditional online identity (Davis and Jurgenson, 2014).

Context collapse is an incredibly important concept for understanding how research is shared, received and acted upon. As researchers embrace multiple identities and their work mediates and is mediated across these wider networked spaces, research seeps into previously unknown locations and can act as a starting point for that work to be understood and have an impact. The flip side of the same coin is that this process is emotionally exhausting and has potentially negative consequences for individuals. Researchers' words can be misunderstood and their personal lives can be exposed to an undefined and unaccountable 'invisible audience' (Davis and Jurgenson, 2014). Deciding what platform to use and why should certainly involve a careful consideration of the inevitable blurring of boundaries in networked online spaces.

One way to cope with the overwhelming idea of context collapse is through Daniel Miller et al.'s (2016) equally fruitful theory for understanding how people interact online: 'scalable sociality', which is discussed throughout the 'Why We Post' book series and at length in *How the World Changed Social Media*. Based on these studies, Miller et al. argue that individuals interact with social media as a whole, as well as with specific platforms, on a spectrum or scale of sociality, depending on the degree of privacy and the size of the group they are looking to reach (2016: 3–8). This spectrum is useful for understanding how people interact on and with social media, but also how we understand social media themselves as dynamic, shifting spaces.

For example, if we consider our fictitious researcher Kira from Chapter 1, she uses social media for a number of different reasons across the Research Lifecycle – to connect with academic colleagues, to share her latest research and to present her research in ways that can be understood by non-academic audiences, etc. These different purposes may mean that some of Kira's social media messages are seen by unintended audiences, which may involve a degree of context collapse. This isn't necessarily problematic but worth considering further. How is Kira presenting her expertise? Who would be her target audiences? With each interaction she may be asking herself what platform suits her needs. If her content is a journal article or a blog post and she is looking to reach as wide a range of researchers as possible on a platform she frequently engages with, then Twitter is a good fit. But say she was looking to have a more private conversation with a discrete number of research participants regarding their thoughts on policy recommendations based on the research? Whatsapp or even a private Facebook group may be a more appropriate platform to encourage people to voice their opinions freely. In communicating your own research-based messages, scalable sociality may prove a helpful way to think through the audience(s) you wish to reach and the online identity(ies) you can cultivate in order to reach them.

Making distinctions between public and private, open and closed, and permanent and ephemeral may help shed light on how to approach various platforms, but in reality, these characteristics often overlap and blur.

7.1.4   Online listening as a deliberate social media strategy

As we've seen in previous chapters on blogging, data visualisations, podcasts and photography and video, there is a wide variety of formats, tools and platforms out there for communicating research, so it is clear that a one-strategy-fits-all approach will not be suitable. But there is one metaphor that can be applied across platforms that can be useful for identifying a more qualitative understanding of effective digital engagement, and that is online listening.

With the diffuse and ever-expanding role of digital media in our daily lives, attention has become a limited resource. Dave Beer has noted the particular influence this will have on academic work, framing the issue in terms of social media's politics of circulation (Beer, 2013). If we don't understand the politics that define contemporary media cultures, then we may also find that researcher practice is reshaped in unforeseen ways. And, more pragmatically, how are researchers themselves coping with all the din? Part of a successful digital strategy for research communication must be dealing head-on with the added social media 'noise' and shaping our responses to this information in productive directions.

Rather than an unmanageable deluge of information towering over passive actors, Kate Crawford helpfully frames the online attention economy of social media in terms of modes of listening:

> The concept of listening ... invokes the more dynamic process of online attention, and suggests that it is an embedded part of networked engagement – a necessary corollary to having a 'voice'. ... Moreover, as a metaphor for attending to discussions and debates online, listening more usefully captures the experience that many Internet users have. It reflects the fact that everyone moves between the states of listening and disclosing online; both are necessary and both are forms of participation. A consideration of listening practices allows for a more acute assessment of online engagement. (Crawford, 2009: 527–528)

Crawford focuses her analysis on Twitter in particular, where different forms of listening practices yield different responses from the social media user, who is both a 'listened to and listening subject' (2009: 533). She argues there is great value in understanding these listening practices as it helps facilitate an understanding of 'the nuances of connection and communication that these spaces afford' (2009: 533). Combining Beer's and Crawford's approaches, engagement online is less about measuring the broadcast (which we will explore below) and more about one's ability to listen in different ways to the different circulation patterns occurring in online spaces.

These kinds of qualitative approaches to digital engagement can yield a more complete understanding of how research is shared, recognised and valued on social media channels. Online listening can be a more passive experience, also known as 'lurking', which can be an incredibly valuable process when you are just starting out on social media, but can also be more active through more direct responses and shares of what other users are contributing. Another active way to set up online listening channels is to deliberately construct your online networks to ensure you are hearing relevant and diverse perspectives. Certain platforms are more conducive to listening than others.

Through hashtags and lists, Twitter is a great tool for tapping into a wide variety of themed conversations. There are many free platforms out there for helping to organise these conversations, like Tweetdeck and Hootsuite, and free social media analysis programs, like FollowerWonk, can help you monitor and analyse your network for better understanding of the strengths and gaps in your network.

What constitutes successful networked listening and engagement? What actually counts in online scholarly interactions beyond that which can be counted? In many ways, these questions are entirely subjective – what counts for some researchers in certain fields, locations and career stages will not matter much for others. No algorithm or network analysis is going to have all the answers for how to get the most out of your social media channels. In order to make sense of all the social media noise, sometimes it is best to just sit back and listen.

7.2   MEASURING DISSEMINATION, ENGAGEMENT AND IMPACT

The key to successful online engagement is to learn what works. This section will explore a range of indicators and analytics tools to help individuals and research teams get a handle on how to measure their social media output.

In conversations about social media and blogging that we, as co-authors, have had over the years, the last three elements of the Research Lifecycle – dissemination, engagement and impact – are the three that come up most frequently, so we have chosen to frame this section around these aspects. But it may be useful before we begin to briefly distinguish again what we mean by each of them:

Dissemination – by this we mean wider awareness of academic work, that is, how many people, academic and non-academic, are coming across your work and what is being done to share and spread this work more widely. To put it another way, we are talking about eyeballs on the page here, but also about targeting different kinds of audiences than you would normally reach.

Engagement is focused on the conversations and meaningful encounters that can occur online and in person as a result of your dissemination efforts and the two-way exchange that is occurring as part of this conversation. Really, engagement is about building effective partnerships. Public engagement is about building effective partnerships with non-academic communities, but engagement can also be about two-way exchange with academic partners.

Impact can be seen in many different ways but we understand it as an 'occasion of influence' or those demonstrable changes resulting from either the dissemination or public engagement activities, or some mixture of the two. In order for research to make an impact, at some point various actors will have had to come across your work. This may come from one-to-one or one-to-many interactions.

Of course, there is actually considerable overlap between and across these terms. But in order to measure the effectiveness of your social media strategy, you must first clearly articulate your aims and means.

7.2.1 Overview of measuring social media in research settings

In the last five years there has been a growing body of literature on how to measure the effectiveness of social media strategies for business (Hanna et al., 2011), advertising (Evans, 2012), civic participation (Obar et al., 2012) and even student learning (Rutherford, 2010), but measuring social media for research purposes has yet to be fully explored. There is certainly useful overlap from how other contexts understand social media metrics, but in order for research aims and values to be fully realised, a different approach to measurement should be cultivated. Our aim in this section is first to explore the value of the research process and the Research Lifecycle and then look at how social media fit within these values. Finally, we identify what types of measurement might come into play from there. By honing in on the specific use and value of social media interactions for distinct research purposes, we argue that a more effective knowledge base can be established to keep tabs on project progress, reports for funders and professional development.

Social media are clearly being used in a variety of ways with many different aims and purposes across research landscapes. Because these types of experiments in networked collaboration and communication are still relatively new, there is a general lack of understanding on how to measure their success in relation to core research aims. Many of the examples and approaches to social media in research contexts mentioned in the previous chapters have developed organically over time, but some were more immediate responses to pressing research needs. With such diversity, there is no one successful digital engagement strategy, but narrowing in on some key aspects related to wider communication and measures of influence will help to clarify a starting point.

In addition to open, distributed forms of knowledge sharing and public engagement, an effective social media strategy also contributes to the academy's understanding of academic influence and the impact of research. 'Influence' is a messy concept, perniciously vague about what actually matters and equally unclear over its subjective determination. Vyacheslav Polonski (2015) offers a succinct definition: 'social influence is the ability to drive causal behavioural change'. This definition, while still all-embracing, helps to point us towards a directed action for influence to have occurred.

Academia is no stranger to using metrics to inform and direct research activities. From bibliometrics that track citations of peer-reviewed work, to assessment exercises that rank academic outputs, to the myriad university rankings that continue to shape university strategy and leadership decision-making, quantitative and qualitative indicators already play a significant role in Higher Education. In 2014, the Higher Education Funding Council for England commissioned an independent report to look at the role of metrics in research assessment. The report (Wilsdon et al., 2015) highlights that while there are great opportunities for the research community to embrace some use of metrics, an over-reliance on poorly-designed indicators will have drastically negative consequences for the sector. Rather, 'responsible metrics' recognise that neither a piece of research nor an individual researcher should be reduced to a simple number.

Beyond academia, there is much to be gained by a cautious approach to measurement. In policymaking, for example, the unwanted effects of targets have long been recognised, and are best encapsulated by British economist Charles Goodhart (Goodhart's Law paraphrased best here by Marilyn Strathern): 'When a measure becomes a target, it ceases to be a good measure' (Strathern, 1997). Metrics can help to capture patterns of influence and broaden our understanding of how content is being used, communicated and acted upon. But by merely identifying a metric, we are all making implicit and explicit judgements on who the content is for and its potential impact.

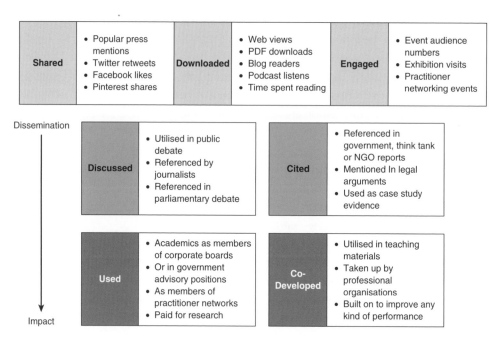

Figure 7.5   Examples of types of impact metrics tracking how research has been used

Credit: Wilsdon, J., et al. (2015). The Metric Tide: Report of the Independent Review of the Role of Metrics in Research Assessment and Management. DOI: 10.13140/RG.2.1.4929.1363 Open Government Licence 2.0: www.nationalarchives.gov.uk/doc/open-government-licence/version/2

These judgements may or may not be sound and could also limit our future understanding of impact and value (Tinkler, 2015). But at this point it seems unwise to remain ignorant of such impact metrics, both the helpful and unhelpful ones, as they are currently playing a significant role in career development and researcher understanding. Figure 7.5 looks at the types of impact metrics identified in 'The metric tide' report (Wilsdon et al., 2015).

While the Impact Factor has been mentioned as a particularly egregious example of an indicator gone wrong, there is a growing movement in the research community devoted to the identification of useful indicators of influence. The Alternative Metrics community, or altmetrics for short, has been instrumental in recent years in furthering the debate about how metrics, including social media metrics, may be of greater use for the research community. These indicators are largely at the research article level (downloads, citations, html pageviews, etc.) and also are increasingly looking to incorporate social media metrics (number of tweets about a journal article, Reddit discussions, blog posts that link back to primary research, etc.). Altmetrics face the same issues that more established indicators face in that as soon as you identify a metric to follow, you are in some way changing the game, but their value is in their breadth of options. Searching for the many ways your work is accessed, used and remixed in different settings and by different audiences should be a continual process.

7.2.2 Qualitative and quantitative indicators

Here we provide a brief overview of qualitative and quantitative indicators of research which may be useful in the evaluation of your own projects. This list is compiled from a range of academic, nonprofit and more traditional media sources. Though distinct differences are apparent in each setting, these sectors are all interested in measuring the impact of engagement, dissemination and impact of communication efforts. Rather than differentiating between the sectors or between the more qualitative and quantitative approaches, we have chosen to present this as one list to encourage further cross-sector investigation of these processes.

Dissemination indicators:

- Reach of media content
- Listens or views of content on external sites
- Number of downloads or pageviews on project website
- Growth of followers on social media
- Retweets or shares from other influential social media accounts
- Mainstream media mentions (with estimated audience figures)

- Reposts on external sites
- Percentage of web traffic from social media
- Newsletter subscribers

EXAMPLE OF DISSEMINATION INDICATOR

According to Google Analytics, your research blog post on low-carbon options for regional development has been viewed by over 10,000 people and inbound links show your work has been mentioned on local government webpages, in environmental charity reports and by several green small businesses in other regions.

Engagement indicators:

- Range of content targeted at multiple target groups
- Likes, shares and comments on social media
- Mentions on social media from research beneficiaries
- Number of comments received
- Other blogs or websites linking to your content
- Time spent on page compared with other media
- Case studies of external partnerships
- Workshop attendance with diverse range of actors
- Sentiment analysis of social media related to research product

EXAMPLE OF ENGAGEMENT INDICATOR

Following a small workshop with small business owners to discuss some emerging opportunities for low-carbon products, small businesses have been offering feedback on your research questions. The team is using this feedback to make an infographic aimed at the wider regional development sector.

Impact indicators:

- Reuse of data or software in another setting
- Invitations to speak at conferences and workshops
- Policy- or decision-makers mentioning research in relation to approach
- Industry partnerships or spin-offs based on research
- New collaborations emerging from research communication
- Effects on local/global community

EXAMPLE OF IMPACT INDICATOR

After several rounds of consultations, the UK's Department for Business, Energy and Industrial Strategy has decided to retire old power plants instead of investing in their refurbishment. The initial workshops with small businesses on their low-carbon portfolio led to a collaboration and the eventual commercialisation of a new biorenewable product (£4.2 million initial investment). Another research team has used the original data for its research looking at taxation mechanisms for low-carbon growth.

7.2.3 What funders want

Demonstrating impact and engagement is now an established part of research assessment and is apparent in messages from research funders worldwide. Many initiatives over the years have focused on the impact of government-funded development research in the Global South, such as Research to Action, Latin American Council on Social Sciences (CLACSO) and Development Research Uptake in Sub-Saharan Africa (DRUSSA). Research funders like the Department for International Development (UK) and the International Development Research Centre (Canada) have been bringing together knowledge intermediaries and research institutes to discuss more effective research communication for years (Barnard et al., 2007). But in other research areas, with the growing importance of two-way exchange and impact, research councils are more and more highlighting the processes and importance of research communication.

The UK was the first country to formally link impact outcomes to *all* government-funded research with its 2014 Research Excellence Framework (Shortt et al., 2016). Impact assessment and the so-called 'impact agenda' more generally is now often associated with UK research and its particular brand of public-sector accountability. But a 2016 report published by Digital Science, 'The societal and economic impacts of academic research: International perspectives on good practice and managing evidence', attests to the increasingly global phenomenon of impact assessment, noting emerging assessment frameworks in Ireland, Australia and Sweden.

As mentioned in the previous sections, a flexible approach to measuring social media is necessary to inform a broad understanding of potential impacts of your project. This can be equally valuable when it comes to communicating this approach in funding applications and reporting documents. Most funders provide some sort of guidance on what they are looking for in relation to communication (see the ESRC Research Impact toolkit; Involve National Institute for Health Research (NIHR) guidance on communicating with research participants through social media; and the Wellcome Trust's public engagement strategy planning advice). The US's National Institutes of Health even has *A Checklist for Communicating Science and Health Research to the Public* for its researchers (NIH, 2016).

Funders are also looking at the new emerging metrics and rolling out strategies to monitor outcomes of research. In a journal article on the potential of alternative metrics to inform funders about research impact, representatives from the Wellcome Trust write: 'altmetrics offer research funders greater intelligence regarding the use and reuse of research, both among traditional academic audiences and stakeholders outside of academia' (Dinsmore et al., 2014). While conventional strategies related to assessing scholarly merit, such as citation counts and peer review, still dominate impact assessment, alternative impact metrics and the online and offline communication efforts that facilitate impact are sure to play a bigger role among funders in years to come.

Central to the value of metrics in grant applications is the competitive advantage for researchers with solid evidence of their research impact. When reviewing grant applications, research funders are looking to ensure the researchers are qualified to investigate pressing questions, can conduct rigorous work and will competently carry through the projected aims. Alternative metrics and indicators of previous success can provide a solid basis for these applications. In *The Research Funding Toolkit*, authors Jacqueline Aldridge and Andrew Derrington (2012: 6) look at why this information is so important and how researchers can pull this information together in a convincing way:

> The most important source of information on your capabilities is your personal track record. Evidence of your previous research performance helps the grants' committee and the referees predict whether you are capable of delivering the proposed programme of research and its outputs. In simple terms, if you have done it before, they will trust you to do it again. If you haven't done it before, then you will have to convince them that you have the ability to do it for the first time.

In such a competitive funding landscape, Aldridge and Derrington (2012: 120) also note the importance of an *assert–justify* approach to grant writing, which prioritises evidence as a way to convince readers and grant committee members that what you are telling them is backed up.

Documenting and presenting evidence of the impact of previous research is, thus, a valuable tool for today's researcher to master. But due to the vague nature of impact and difficulty of its measurement, as discussed earlier in this chapter, this is not as simple as following a pre-determined list of metrics. Funders are extremely reticent about suggesting specific metrics as this may lead to unintended consequences on researcher behaviour and gaming the system (Curry, 2014). So while such dissemination and impact metrics could be incredibly useful for researchers in their grant applications, determining which metrics *should* be used regarding your project may be a more difficult question to answer. New tools like Altmetric and ImpactStory are looking to help provide a range of useful information, but data accuracy and consistency still remain an issue (Gordon et al., 2015). Indeed, one of the recommendations of 'The metric tide:

Report of the independent review of the role of metrics in research assessment and management' (Wilsdon et al., 2015) identifies the underlying infrastructure of these metrics as an issue of great importance: 'Indicators can only meet their potential if they are underpinned by an open and interoperable data infrastructure' (Wilsdon et al., 2015). So far, there has been little institutional engagement over who owns the metric data, how the data are structured and what responsible use of metrics looks like.

With the introduction of social media and the privileging of these new metrics in research assessment, there has been some worry that researchers will feel forced to over-emphasise the communication of research at the expense of the research itself. On the whole, social media offers considerable opportunities for the research environment, but misplaced objectives regarding social media use are still a legitimate concern that research funders and institutions should take seriously. If researchers and knowledge workers do not understand why they are using social media (other than because of some top-down requirement), this will undoubtedly have negative consequences for research and will also lead to considerable wasted time and effort. The communication of research can only ever be as good as the research we are communicating. But communication of research and its assessment and measurement is also not an activity so neatly separated from the research process itself. These activities coexist in often coordinated but sometimes conflicted ways. Cautious attitudes regarding social and digital media make this very apparent. We conclude this section with a reflection on how researchers and institutions can navigate the risks of online visibility and social media use in the research environment to ensure productive and positive directions for researchers and research users.

7.3 NAVIGATING THE RISKS OF ONLINE VISIBILITY

We have spent most of this book telling you how using social media in the research process can go right, but we also need to acknowledge that it can, and often does, go wrong. It is important to consider the costs of social media's introduction to the research environment.

The most glaring risk related to social media engagement is online harassment and abuse, aimed disproportionately at women and people of colour. As a high-profile researcher in the public eye, Professor Mary Beard has written about the topic over the years in her blog *A Don's Life*. In a 2013 post entitled 'Internet fury: or having your anatomy dissected online', Beard shares a snapshot of the messages she received after one of her appearances on BBC's *Question Time*. She writes, 'the misogyny here is truly gobsmacking. ... It would be quite enough to put many women off appearing in public, contributing to political debate, especially as all of this comes up on google' (Beard, 2013). Beard situates this particular brand of vileness on social media not as something unique to anonymous online spaces but as part of a wider cultural tradition of attacking women who appear in the public sphere: 'It doesn't much matter what line of argument you take as a woman. If you venture into traditional male territory, the abuse comes anyway. It's not what you say that prompts it – it's the fact that you are saying it' (Beard, as quoted in Mead, 2014).

In 2016, *The Guardian* commissioned research on its comments section and found that articles written by women attract more abuse and trolling than those written by men (Gardiner et al., 2016). Of the ten most abused writers in their analysis, eight were women and two were black men. Much has been written on the topic, but frustratingly little has actually been done to stop the flurry of vile comments directed at individuals. Though this is not a universal experience for every woman who engages online, the particularly offensive abuse directed at those in the spotlight, combined with the fact that little responsibility seems to be taken by platforms in addressing these issues, makes online public engagement an incredibly frustrating and off-putting space.

The problem of abuse and bullying should be taken seriously, especially by funders and institutions looking to promote the wider public presence of researchers and their work. Especially for early-career researchers who are already in a position of vulnerability in relation to job security, taking on the burden of an online presence without adequate training or support mechanisms could have very damaging effects for the researcher. Reflecting on a panel that looked at the challenges and risks of being a public intellectual, Audra Mitchell (2013) writes: 'Some individuals may be able to shrug off disturbing comments or even physical threats, but for others this is a deeply damaging experience – and it is most certainly not part of "what we sign up for".' Universities, funding councils and other research bodies need to consider their role in shaping a positive research environment and think carefully about the risks and rewards involved in an individual's online presence.

Educator and social network researcher Bonnie Stewart argues that vulnerabilities are created by the combination of two factors central to the experience of social media: on the one hand, the overarching phenomenon of callout culture, which for researchers means they as individuals or their work could be 'caught out' and misrepresented via unsubstantiated critiques; and second, what Stewart (2016) describes as 'the collapse of oral and literate cultural habits and audience expectations', which for researchers means things they say in passing can be taken out of context and spread to a wide, unintended audience. Twitter is an open public space that thrives as an informal platform, but its 'artifacts', all the digital traces which make it up, also follow the properties of online networked content more generally in that they are persistent, replicable, scaleable and searchable (boyd, 2011: 46). When content which is oral in nature (i.e. casual and with a social function) is embedded within a platform supported by these networked properties, new dynamics, behaviours and norms emerge. The differences between the ways speech and work are made public through traditional media and new media present new opportunities but also vulnerabilities.

Given these vulnerabilities, scholars require structures that take these factors into account and can provide support for their digital activities. But are we currently building these structures in our institutions? Institutions are historically, and perhaps even understandably given their scope, risk-averse, but as we have seen above, engagement and scholarship in the 21st century is inherently filled with risk. Speaking at the LSE in 2016, Stewart ended her talk on networked publics on a challenging note for the future of university social media spaces:

> If as an academy we become good at [institutional approaches to digital spaces] then we can use our digital identities to contribute to public spheres in positive ways. If we do not become good at these things I think that we face a great deal of risk. (Stewart, 2016)

There is no easy way to avoid these negative and complicated aspects of social media use. But an awareness of where the vulnerabilities are emerging and how we can support and protect the most vulnerable among us will help alleviate some of the more glaring contradictions of an online research presence.

Another important risk related to social media use is the echo chamber effect. Unpleasant comments and a lack of support with dealing with the vulnerabilities are causes for concern, but another considerable cause for concern is social media's 'dangerous insularity' (Lewis, 2015). Algorithms that drive many social media platforms depend on sentiment analysis where content is filtered and targeted to users based on previous behaviour (van Dijck and Poell, 2013; Bakshy et al., 2015). Furthermore, users' own behaviour on social media (and offline) to follow those we agree with and share content accordingly can provide a misleading portrayal of society at large (Aiello et al., 2012; Boutyline and Willer, 2016). But insularity is not just a concern for Facebook and Twitter users. According to empirical analysis by Gilbert et al. (2009), agreement overwhelmingly outnumbers disagreement on blog comments with little discussion and debate taking place, suggesting blogs are also susceptible to the echo chamber effect. Is social media just another way for researchers to preach to the choir?

The echo chamber issue has been debated in research circles considerably after the UK's 'Brexit' EU referendum where the vast difference of views between those in the research community and the wider public became very apparent (Chater, 2016; Van Reenen, 2016). But beyond political polarisation, social media's echo chamber effect can lull researchers into thinking that a high engagement rate on social media is in fact true engagement. Social media platforms certainly bear some, if not most, of the responsibility for this situation, but as discussed earlier in section 7.1.4 on online listening, researchers can also look to address the echo chamber effect by confronting confirmation bias and constructing engagement around a diversity and plurality of voices.

Our Research Lifecycle Framework is proposed with these risks and complexities in mind. We do not argue that every researcher should or even could adopt a widespread and comprehensive approach to social media use suitable for every research purpose. Rather, through the previous chapters we have looked to equip researchers, knowledge workers, administrators and those working within the research environment with an understanding of how these social activities fit alongside academic practice in order to help researchers feel more confident navigating the different types of content and media most suitable to their needs. And though we acknowledge a few of the major risks above, new risks will continue to emerge.

7.4 CONCLUSIONS

Social media across the Research Lifecycle encompasses far more than just discrete, individual tasks of collaboration, analysis and communication. They aim to capture the truly social nature of scholarship and reflect the full spectrum of activity in the heavily mediated context of the 21st century, from idea generation and co-creation through to research's impact and influence in society. The Lifecycle model of research provides a useful backdrop to explore how social media can be beneficially employed in research settings. Furthermore, it theoretically grounds social media's role in research settings.

Through exploring the fundamental aspects of the modern Research Lifecycle and an extensive look at how to create engaging content, it is clear that research is now and has in many ways always been a social process. Researchers are already participating in and employing a number of new media and digital technologies, enabling research and facilitating wider connections with this research. But the conditions and contexts of sociality in the digital world are embedded with complexity, risk and vulnerabilities, which are equally bound to have implications for research.

In an episode of the popular technology podcast *Reply All*, host Alex Goldman argues: 'we know that a piece of technology can't make people better or worse. … All technology can do is give us new options for how to behave' (see also the epigram of Chapter 2). An essential theme running through each of our how-to chapters is that any one piece of technology can't make research better or worse, but it can help researchers behave in new ways. Social media tools and technology offer new options for researchers to act in more social ways, thereby realising the social potential of research. But given the complexity in this space, it is important to consider and reconsider how the application of digital technology is itself shaping research. Heightened awareness and sensitivity to the dynamics of these networked spaces are required.

These chapters have explored this social potential of research by looking at how researchers are currently using blogs, podcasts, data visualisations, videos and photography. We have also looked to provide an in-depth look at how these tools can be employed throughout the Research Lifecycle. We hope that our Research Lifecycle Framework, the extensive overviews of different types of new media content and the numerous case studies will inspire further experimentation with social media and the social nature of research, but we also recognise that social media are not merely a one-way conduit for research. Social media have the potential to have profound implications, positive and negative, for how research is conducted and constructed. No one should be forced to use them, but given the possibilities across the Research Lifecycle, and in particular for research engagement, dissemination and impact, we are confident that further experimentation will benefit the research community and invigorate the role of research in society.

7.5   FURTHER READING

Bilder, G., J. Lin and C. Neylon (2015) 'Principles for open scholarly infrastructures.' Available at: https://figshare.com/articles/Principles_for_Open_Scholarly_Infrastructures_v1/1314859 [Accessed 14 November 2016].
A set of design principles for the scholarly community to consider to ensure stable, sustainable and trusted tools, platforms and other facilities of scholarly work.

Crawford, K. (2009) 'Following you: Disciplines of listening in social media.' *Continuum*, 23(4): 525–535.
This article helpfully interrogates what it means to be an active participant on social media. Listening is used as a metaphor for paying attention online, challenging the notion that 'lurking' behaviour is less active than, for example, users who tweet and share new material often.

Davis, J. and N. Jurgenson (2013) 'Context collapse: A literature review.' *Cyborgology*. January 10. Available at: https://thesocietypages.org/cyborgology/2013/01/10/context-collapse-a-literature-review [Accessed 14 November 2016].
Background and a list of resources on the structure of networked publics and the theory of context collapse.

Miah, A. (2016) 'The A to Z of social media for academia.' *Times Higher Education*. August 18. Available at: www.timeshighereducation.com/a-z-social-media [Accessed 14 November 2016].
An exhaustive and regularly updated list of social media platforms that have been used for research purposes.

Stewart, B. (2016) 'Collapsed publics: Orality, literacy, and vulnerability in academic Twitter.' *Journal of Applied Social Theory*, 1(1). Available at: http://socialtheoryapplied.com/journal/jast/article/view/33/9 [Accessed 14 November 2016].
An ethnographic study of academic interactions on Twitter and an in-depth look at how context collapse occurs in academic environments.

REFERENCES

113 Cong. Rec. (2014a) 160(42) S1599 (Statement of Sen. John Barrasso)

113 Cong. Rec. (2014b) 160(135) E1475 (Statement of Rep. Keith Ellison)

114 Cong. Rec. (2015) 161(108) S5002 (Statement of Sen. Patrick Leahy)

Ackroyd, P. (2001) *London: The Biography*. London: Vintage.

Aiello, L. M., Barrat, A., Schifanella, R., Cattuto, C., Markines, B. and Menczer, F. (2012) 'Friendship prediction and homophily in social media.' *ACM Trans. Web*, 6(2): 9: 1–9:33.

AIR (Association of Independents in Radio) (2016) 'The Alex Kapelman guide to the creation of a dope project: 9 tips to keep indie podcasters happy, sane and moving forward.' *AIR website*. February 15. Available at: http://airmedia.org/the-alex-kapelman-guide-to-the-creation-of-a-dope-project-9-tips-to-keep-indie-podcasters-happy-sane-and-moving-forward [Accessed 24 July 2016].

Alcorn, S. (2014) 'Is this thing on?' *Digg*. January 15. Available at: http://digg.com/originals/why-audio-never-goes-viral [Accessed 20 March 2016].

Aldred, R. (2014) *Benefits of Investing in Cycling* [Report]. British Cycling. Available at: www.britishcycling.org.uk/zuvvi/media/bc_files/campaigning/BENEFITS_OF_INVESTING_IN_CYCLING_DIGI_FINAL.pdf [Accessed 1 February 2016].

Aldrich, J. H., Gibson, R.K., Cantijoch, M. and Konitzer, T. (2015) 'Getting out the vote in the social media era: Are digital tools changing the extent, nature and impact of party contacting in elections?' *Party Politics*, 22(2): 165–178.

Aldridge, J. and Derrington, A.M. (2012) *The Research Funding Toolkit: How to Plan and Write Successful Grant Applications*. London and Thousand Oaks, CA: Sage.

Alexander, B. (2006) 'Web 2.0: A new wave of innovation for teaching and learning?' *EDUCAUSE Review*, 41(2): 32–44.

Allen, H.G., Stanton, T.R., Di Pietro, F. and Moseley, G.L. (2013) 'Social media release increases dissemination of original articles in the clinical pain sciences.' *PLoS One*, 8(7): e68914.

Alperin, J. P. (2014) 'Altmetrics could enable scholarship from developing countries to receive due recognition.' *LSE Impact Blogs*. March 10. Available at: http://blogs.lse.ac.uk/impactofsocialsciences/2014/03/10/altmetrics-for-developing-regions [Accessed 14 November 2016].

Alperin, J. P., Fischman, G. E. and Willinsky, J. (2012) 'Estrategias de comunicación académica en universidades de investigación intensiva de América Latina.' *Educación Superior y Sociedad*, 16(2). Available at: http://ess.iesalc.unesco.org.ve/index.php/ess/article/view/409 [Accessed 14 November 2016].

Amar, Z. (2015) 'Like Instagram: It'll do wonders for your charity.' *The Guardian*. April 21. Available at: www.theguardian.com/voluntary-sector-network/2015/apr/21/charity-instagram-tips-fundraising-campaign [Accessed 14 November 2016].

Aminuddin, I. (2016) *The Ice-cream Story: PrettyEveel Adventures.* Available at: http://sites.northwestern.edu/eveeleva/2016/02/12/the-ice-cream-story [Accessed 14 August 2016].

Anderson, C. (2008) *The Long Tail: Why the Future of Business is Selling Less of More.* Revised edition. New York: Hyperion.

Anwar, N.H., Mustafa, D., Sawas, A. and Malik, S. (2016) *Gender and Violence in Urban Pakistan.* March. Available at: www.idrc.ca/en/project/gender-and-violence-urban-pakistan [Accessed 14 November 2016].

Arana, G. (2015) '*538*'s Nate Silver to *Vox*: Quit jacking our charts.' *The Huffington Post.* Available at: www.huffingtonpost.com/2015/04/13/nate-silver-vox-chart_n_7056154.html [Accessed 14 August 2016].

ARL (2016) 'Scholarly communication | Association of Research Libraries® | ARL®.' Available at: www.arl.org/focus-areas/scholarly-communication [Accessed 2 May 2016].

Arrington, M. (2006) *Odeo Releases Twttr.* [online] TechCrunch. Available at: http://techcrunch.com/2006/07/15/is-twttr-interesting [Accessed 2 May 2016].

Avey, P.C. and Desch, M.C. (2014) 'What do policymakers want from us? Results of a survey of current and former senior national security decision makers.' *International Studies Quarterly*, 58(2): 227–246.

Avirgan, J. (2016) *Postcards.* [Podcast] FiveThirtyEight. March 10. Available at: http://fivethirtyeight.com/features/dear-data-and-fivethirtyeight-want-you-to-visualize-your-podcast-habits [Accessed 14 November 2016].

Babini, D. (2013) 'Open access initiatives in the Global South affirm the lasting value of a shared scholarly communications system.' *LSE Impact of Social Sciences*. October 23. Available at: http://blogs.lse.ac.uk/impactofsocialsciences/2013/10/23/global-south-open-access-initiatives [Accessed 14 November 2016].

Back, L. and Bull, M. (eds) (2003) *The Auditory Culture Reader.* Oxford: Berg.

Badger, G. (2007) *The Genius of Photography: How Photography has Changed Our Lives.* London: Quadrille.

Bakhshi, S., Shamma, D.A. Kennedy, L. and Gilbert, E. (2015) 'Why we filter our photos and how it impacts engagement.' In *Ninth International AAAI Conference on Web and Social Media*. April. Available at: http://comp.social.gatech.edu/papers/icwsm15.why.bakhshi.pdf [Accessed 14 November 2016].

Bakshy, E., Messing, S. and Adamic, L.A. (2015) 'Exposure to ideologically diverse news and opinion on Facebook.' *Science*, 348(6239): 1130–1132.

Bales, K. (1999) 'Popular reactions to sociological research: The case of Charles Booth.' *Sociology*, 33(February): 153–168.

Barlow, A. (2007) *The Rise of the Blogosphere*. Westport, CT: Greenwood Publishing Group.

Barlow, A. (2008) *Blogging America: The New Public Sphere*. Westport, CT: Greenwood Publishing Group.

Barnard, G., Carlile, L. and Ray, D.B. (2007) 'Maximising the impact of development research: How can funders encourage more effective research communication?' *Institute of Development Studies*. Working Paper Series. Available at: www.ids.ac.uk/files/research_comms_final_report_max_impactMarch07.pdf [Accessed 14 November 2016].

Basara, L.R. and Juergens, J.P. (1994) 'Patient package insert readability and design.' *American Pharmacy*, 34(8): 48–53.

Bastow, S., Dunleavy, P. and Tinkler, J. (2014) *The Impact of the Social Sciences: How Academics and their Research Make a Difference*. London: Sage.

Bauer, M.W. (2009) 'The evolution of public understanding of science: Discourse and comparative evidence.' *Science, Technology & Society*, 14(2): 221–240.

Baym, N.K. and boyd, d. (2012) 'Socially mediated publicness: An introduction.' *Journal of Broadcasting & Electronic Media*, 56(3): 320–329.

Beard, M. (2013) 'Internet fury: Or having your anatomy dissected online.' *Blogpost on a Don's Life*. January 27. Available at: http://timesonline.typepad.com/dons_life/2013/01/internet-fury.html [Accessed 14 November 2016].

Beashel, A. (2014) 'Email marketing vs social media: Are you focusing on the wrong channel?' *Campaign Monitor Blog*. Available at: https://www.campaignmonitor.com/blog/email-marketing/2014/07/email-marketing-vs-social-media [Accessed 1 November 2016].

Beaver, D.D. (2001) 'Reflections on scientific collaboration (and its study): Past, present, and future.' *Scientometrics*, 52(3): 365–377.

Becker, H.S. (1995) 'Visual sociology, documentary photography, and photojournalism: It's (almost) all a matter of context.' *Visual Studies*, 10(1): 5–14.

Beer, D. (2009) 'Power through the algorithm? Participatory web cultures and the technological unconscious.' *New Media & Society*, 11(6): 985–1002.

Beer, D. (2013) 'Social media's politics of circulation have profound implications for how academic knowledge is discovered and produced.' *Impact of Social Sciences Blog*. July 29. Available at: http://blogs.lse.ac.uk/impactofsocialsciences/2013/07/29/academic-knowledge-and-the-politics-of-circulation [Accessed 2 May 2016].

Benson, V. (ed.) (2014) *Cutting-Edge Technologies and Social Media Use in Higher Education*. Hershey, PA: IGI Global.

Bentley, P. and Kyvik, S. (2010) 'Academic staff and public communication: A survey of popular science publishing across 13 countries.' *Public Understanding of Science*, 20(1): 48–63.

Bergstrom, T. (2014) 'Secrets of journal subscription prices: For-profit publishers charge libraries two to three times more than non-profits.' *Impact Blog. LSE.* December 8. Available at: http://blogs.lse.ac.uk/impactofsocialsciences/2014/08/12/secrets-of-the-big-deal-journal-pricing [Accessed 27 April 2016].

Berry, R. (2005) 'Will the iPod kill the radio star? Profiling podcasting as radio.' *Convergence*, 12(2): 143–162.

Berry, R. (2015) 'A golden age of podcasting? Evaluating *Serial* in the context of podcast histories.' *Journal of Radio and Audio Media*, 22(2): 170–178.

Bestley, R. (2013) *Visual Rhetoric.* Available at: http://visualrhetoric.ac.uk/about-2 [Accessed 1 February 2016].

Bignante, E. (2010) 'The use of photo-elicitation in field research: Exploring Maasai representations and use of natural resources.' *EchoGéo*, 11. Available at: https://echogeo.revues.org/11622?lang=en [Accessed 14 November 2016].

Bik, H.M. and Goldstein, M.C. (2013) 'An introduction to social media for scientists.' *PLoS Biol*, 11(4), p.e1001535.

Bilder, G., Lin, J. and Neylon, C. (2015) 'Principles for open scholarly infrastructures-V1.' Available at: https://figshare.com/articles/Principles_for_Open_Scholarly_Infrastructures_v1/1314859 [Accessed 1 November 2016].

Björk, B.C. and Solomon, D. (2013) 'The publishing delay in scholarly peer-reviewed journals.' *Journal of Informetrics*, 7(4): 914–923.

Bly, R.W. (2007) *Blog Schmog: The Truth about What Blogs Can (and Can't) Do for Your Business.* Nashville, TN: Thomas Nelson Inc.

Blyth, M. (2012) 'Five minutes with Mark Blyth: "Turn it into things people can understand, let go of the academese, and people will engage."' *LSE Impact Blog.* March 9. Available at: http://blogs.lse.ac.uk/impactofsocialsciences/2012/03/09/five-minutes-with-mark-blyth [Accessed 14 November 2016].

Booth, C. (1889) *Life and Labour of the People: First Series – Poverty (i) East, Central and South London.* London: Macmillan. (Republished 1969.)

Booth, C. (1902a) *Life and Labour of the People: First Series – Poverty (ii) Streets and Population Classified.* London: Macmillan. (Republished 1969.)

Booth, C. (1902b) *Life and Labour of the People in London: Final Volume – Notes on Social Influences and Conclusions.* London: Macmillan.

Boulay, R. (2013) 'Designing and developing online materials for molecular biology: Building online programs for science.' *International Journal of Design Education*, 6(3): 53.

Bourdieu, P. (1990) *Photography: A Middle-brow Art.* Cambridge: Polity Press.

Boutyline, A. and Willer, R. (2016) 'The social structure of political echo chambers: Variation in ideological homophily in online networks.' *Political Psychology*, April. doi:10.1111/pops.12337.

Box, G.E.P and Draper, N. R. (1987) *Emperical Model Building and Response Surfaces*. New York: John Wiley & Sons.

boyd, d. (2007) 'Social network sites: Public, private, or what?' *Knowledge Tree*, May 13. Available at: www.danah.org/papers/KnowledgeTree.pdf [Accessed 14 November 2016].

boyd, d. (2010) 'Facebook is a utility: Utilities get regulated.' *Zephoria.org*, May 15. Available at: www.zephoria.org/thoughts/archives/2010/05/15/facebook-is-a-utility-utilities-get-regulated.html [Accessed 14 November 2016].

boyd, d. (2011) 'Social network sites as networked publics: Affordances, dynamics, and implications.' In Z. Papacharissi (ed.), *A Networked Self: Identity, Community, and Culture on Social Network Sites*. London: Routledge.

boyd, d. and Crawford, K. (2012) 'Critical questions for Big Data.' *Information, Communication & Society*, 15(5): 662–679.

boyd, d. and Ellison, N.B. (2007) 'Social network sites: Definition, history, and scholarship.' *Journal of Computer-Mediated Communication*, 13(1): 210–230.

Boyer, E.L. (1997) *Scholarship Reconsidered: Priorities of the Professoriate*. First Edition. Princeton, NJ: Jossey Bass.

Brabazon, T. (ed.) (2012) *Digital Dialogues and Community 2.0: After Avatars, Trolls and Puppets*. Oxford: Chandos Publishing.

Brasseur, L. (2005) 'Florence Nightingale's visual rhetoric in the rose diagrams.' *Technical Communication Quarterly*, 14(2): 161–182.

Brecht, B. (1932) 'The radio as an apparatus of communication.' In Neil Strauss (ed.), *Radiotext (E)*. New York: Columbia University Press, 1993, p. 15.

Brienza, C. (2014) 'Paying twice or paying thrice? Open access publishing in a global system of scholarly knowledge production and consumption.' *LSE Impact Blog*. January 30. Available at: http://blogs.lse.ac.uk/impactofsocialsciences/2014/01/30/paying-twice-or-paying-thrice-brienza [Accessed 14 August 2016].

Brown, I. (2014) 'Social media surveillance.' *The International Encyclopedia of Digital Communication and Society*, 1–7.

Brugha, R. and Varvasovszky, Z. (2000) 'Stakeholder analysis: A review.' *Health Policy and Planning*, 15(3): 239–246.

Brumley, C. (2013) 'LSERB Editor's column extra II: Nigel Warburton.' *LSE Review of Books* [Podcast]. February 17. Available at: http://blogs.lse.ac.uk/lsereviewofbooks/2013/02/17/editors-column-the-simple-guide-to-academic-podcasting [Accessed 14 November 2016].

Brumley, C. (2014) 'Risky research.' Royal Geographical Society Annual Conference. Slideshow. August 28.

Bruns, A. and Jacobs, J. (2006) 'Introduction.' *Uses of Blogs*, 38: 1–9.

Burchall, K. (2015) 'Factors affecting public engagement by researchers.' Available at: https://wellcome.ac.uk/sites/default/files/wtp060036.pdf [Accessed 1 February 2016].

Burdick, A. and Willis, H. (2011) 'Digital learning, digital scholarship and design thinking.' *Design Studies*, 32(6): 546–556.

Burke, C. (2009) 'Isotype: Representing Social Facts Pictorially.' *Information Design Journal*, 17(3), 210–221.

Cameron, W.B. (1963) *Informal Sociology: A Casual Introduction to Sociological Thinking*. New York: Random House.

Cammarata, L. (2010) 'Welcoming our littlest visitors to the Smithsonian Air and Space Museum.' Available at: https://airandspace.si.edu/stories/editorial/welcoming-our-littlest-visitors-air-and-space-museum [Accessed 14 August 2016].

Carlson, N. (2011) *The Real History of Twitter*. [online] Tech Insider. Available at: www.techinsider.io/how-twitter-was-founded-2011-4 [Accessed 2 May 2016].

Carnall, M. (2014) 'Is Archaeopteryx a bird or not?' *Museums & Collections Blog*. January 23. Available at: http://blogs.ucl.ac.uk/museums/2014/01/23/is-archaeopteryx-a-bird-or-not [Accessed 14 August 2016].

Carr, D. (2014) '*Serial*: Podcasting's first breakout hit sets stage for more.' *The New York Times*. November 24. Available at: www.nytimes.com/2014/11/24/business/media/serial-podcastings-first-breakout-hit-sets-stage-for-more.html?_r=0 [Accessed 24 May 2016].

Carrigan, M. (2016) *Social Media for Academics*. London: Sage.

Carroll, S. (2011) 'How to get tenure at a major research university.' Personal blog. Available at: www.preposterousuniverse.com/blog/2011/2003/2030/how-to-get-tenure-ata-major-research-university [Accessed 14 November 2016].

Cartaldo, C. (2015) 'Fedele paga una messa per "fermare l'invasione islamica"'. *Il Giornale*. Available at: www.ilgiornale.it/news/cronache/fedele-paga-messa-fermare-linvasione-islamica-1179719.html [Accessed 1 May 2016].

Castelvecchi, D. (2015) 'Physics paper sets record with more than 5,000 authors.' *Nature News & Comment*. Available at: www.nature.com/news/physics-paper-sets-record-with-more-than-5-000-authors-1.17567 [Accessed 13 August 2016].

Chater, J. (2016) 'What the EU Referendum result teaches us about the dangers of the echo chamber.' *The New Statesman*. Available at: www.newstatesman.com/2016/07/what-eu-referendum-result-teaches-us-about-dangers-echo-chamber [Accessed 9 August 2016].

China, I. (2015) 'Weibo MAUs reached 222 million in Q3 2015.' *China Internet Watch*. Available at: www.chinainternetwatch.com/15740/weibo-q3-2015 [Accessed 2 May 2016].

Christensen, W. and Suess, R. (1978) 'Hobbyist computerized bulletin board.' *Byte Magazine*, 3(11): 150–158.

Cited (2015) Cited Podcast. Available at: http://citedpodcast.com [Accessed 20 November 2015].

Clarke, H.D., Goodwin, M. and Whiteley, P. (2016) 'Leave was always in the lead: Why the polls got the referendum result wrong'. *LSE British Politics and Policy Blog*. Available at: http://blogs.lse.ac.uk/politicsandpolicy/eu-referendum-polls [Accessed 15 November 2016].

Cogburn, D.L. and Espinoza-Vasquez, F.K. (2011) 'From networked nominee to networked nation: Examining the impact of Web 2.0 and social media on political participation and civic engagement in the 2008 Obama campaign.' *Journal of Political Marketing*, 10(1–2): 189–213.

Cohen, I.B. (1984) 'Florence Nightingale.' *Scientific American*, 250(3): 128–137.

Colleoni, E., Rozza, A. and Arvidsson, A. (2014) 'Echo chamber or public sphere? Predicting political orientation and measuring political homophily in Twitter using Big Data.' *Journal of Communication*, 64(2): 317–332.

Constine, J. (2015) 'Facebook hits 8 billion daily video views, doubling from 4 billion in April.' [Blog post]. *Tech Crunch*. November 4. Available at: https://techcrunch.com/2015/11/04/facebook-video-views [Accessed 14 November 2016].

Cottom, T.M. (2012) 'Risk and ethics in public scholarship.' *University of Venus Blog, Inside Higher Ed.* Available at: www.insidehighered.com/blogs/university-venus/risk-and-ethics-public-scholarship [Accessed 1 January 2016].

Couldry, N. (2015) 'Social media: human life.' *Social Media + Society*, 1(1): 1–2.

Couldry, N. and van Dijck, J. (2015) 'Researching social media as if the social mattered.' *Social Media + Society*, 1(2): 1–7.

Crawford, K. (2009) 'Following you: Disciplines of listening in social media.' *Continuum*, 23(4): 525–535.

Crawford, K. and Gillespie, T. (2014) 'What is a flag for? Social media reporting tools and the vocabulary of complaint.' *New Media & Society*, doi: 10.1177/1461444814543163.

Criado-Perez, C. (2015) 'How social media helped me get Jane Austen on to £10 notes.' *The Guardian*. April 11. Available at: www.theguardian.com/lifeandstyle/2015/apr/11/how-social-media-helped-jane-austen-banknotes-caroline-criado-perez [Accessed 2 May 2016].

Curry, S. (2014) 'Debating the role of metrics in research assessment.' *Reciprocal Space*. October 7. Available at: http://occamstypewriter.org/scurry/2014/10/07/debating-the-role-of-metrics-in-research-assessment [Accessed 7 August 2016].

Curtis, L., Edwards, C., Fraser, K.L., Gudelsky, S., Holmquist, J., Thornton, K. and Sweetser, K.D. (2010) 'Adoption of social media for public relations by nonprofit organizations.' *Public Relations Review*, 36(1): 90–92.

D'Ignazio, C. (2015) 'What would feminist data visualisation look like?' [Blog post]. *MIT Center for Civic Media*. December 20. Available at: https://civic.mit.edu/feminist-data-visualization [Accessed 1 January 2016].

Daniel, P. (2015) 'The secret to 'Serial': An afternoon with Sarah Koenig.' *Huffington Post*. April 16. Available at: www.huffingtonpost.com/patrick-daniel/the-secret-to-serial-an-a_b_6995606.html [Accessed 24 April 2016].

Davis III, C.H., Deil-Amen, R. Rios-Aguilar, C., and Gonzalez Canche, M.S. (2012) *Social Media in Higher Education: A Literature Review and Research Directions*. Report printed by the University of Arizona and Claremont Graduate University. Available at: https://works.bepress.com/hfdavis/2 [Accessed 14 November 2016].

Davis, G.A., Burggraf-Torppa, C., Archer, T.M. and Thomas, J.R. (2007) 'Applied research initiative: Training in the scholarship of engagement.' *Journal of Extension*, 45(2). Available at: www.joe.org/joe/2007april/a2.php [Accessed 1 January 2016].

Davis, J.L. and Jurgenson, N. (2014) 'Context collapse: Theorizing context collusions and collisions.' *Information, Communication & Society*, 17(4): 476–485.

De Roure, D., Jennings, N.R. and Shadbolt, N.R. (2003) 'The semantic grid: A future e-science infrastructure.' In F. Berman, G. Fox and T. Hey (eds), *Grid Computing*. Chichester: John Wiley & Sons. pp. 437–470. Available at: http://onlinelibrary.wiley.com/doi/10.1002/0470867167.ch17/summary [Accessed 2 February 2016].

Denzin, N.K. and Lincoln, Y.S. (2005) *The SAGE Handbook of Qualitative Research*. Third edition. London: Sage.

Diakopoulos, N., Kivran-Swaine, F. and Naaman, M. (2011) 'Playable data: Characterizing the design space of game-y infographics.' *CHI 2011 Proceedings of the SIGCHI Conference on Human Factors in Computing Systems*. Vancouver, BC: Canada. pp. 1717–1726.

DiMeo, N. (2016) 'Finishing hold.' *Memory Palace*. Episode 86. Podcast. April 8. Available at: http://thememorypalace.us/2016/04/finishing-hold [Accessed 14 November 2016].

Dinsmore, A., Allen, L. and Dolby, K. (2014) 'Alternative perspectives on impact: The potential of ALMs and altmetrics to inform funders about research impact.' *PLoS Biol*, 12(11): e1002003.

Direct Marketing Association (2015) 'National client email report 2015.' Available at: http://cdn.emailmonday.com/wp-content/uploads/2015/04/National-client-email-2015-DMA.pdf [Accessed 1 November 2016].

Dolan, J. (2006) 'Why I blog: On the theories and practices of feminist blogging.' *The Feminist Spectator*. October 26. Available at: http://feministspectator.princeton.edu/2006/10/26/why-i-blog-on-the-theories-and-practices-of-feminist-blogging [Accessed 14 August 2016].

Dorling, D., Mitchell, R., Shaw, M., Orford, S. and Smith, G.D. (2000) 'The Ghost of Christmas Past: Health effects of poverty in London in 1896 and 1991.' *BMJ*, 321(7276): 1547–1551.

Doshi, V. (2016) 'Facebook under fire for "censoring" Kashmir-related posts and accounts.' *The Guardian*. July 19. Available at: www.theguardian.com/technology/2016/jul/19/facebook-under-fire-censoring-kashmir-posts-accounts [Accessed 14 August 2016].

Dowse, R., Ramela, T., Barford, K. and Browne, S. (2010) 'Developing visual images for communicating information about antiretroviral side effects to a low-literate population.' *African Journal of AIDS Research*, 9(3): 213–224.

Dredze, M. (2012) 'How social media will change public health.' *Intelligent Systems, IEEE*, 27(4): 81–84.

Dunleavy, P. (2014) 'Shorter, better, faster, free: Blogging changes the nature of academic research, not just how it is communicated.' *Impact of Social Sciences Blog*. Available at: http://blogs.lse.ac.uk/impactofsocialsciences/2014/12/28/shorter-better-faster-free [Accessed 14 August 2016].

Dunleavy, P. and Gilson, C. (2012) 'Five minutes with Patrick Dunleavy and Chris Gilson: Blogging is quite simply, one of the most important things that an academic should be doing right now.' *Impact of Social Sciences Blog*. Available at: http://eprints.lse.ac.uk/51915 [Accessed 14 August 2016].

Dunn, S. and Hedges, M. (2012) 'Crowd-sourcing scoping study: Engaging the crowd with humanities research.' *Arts and Humanities Research Council Report.* Available at: www.ahrc.ac.uk/Funding-Opportunities/Research-funding/Connected-Communities/Scoping-studies-and-reviews/Documents/Crowd Sourcing in the Humanities.pdf [Accessed 22 July 2016].

Durose, C. and Tonkiss, C. (2013) 'Fast scholarship is not always good scholarship: Relevant research requires more than an online presence.' *Impact of Social Sciences Blog.* Available at: http://blogs.lse.ac.uk/impactofsocialsciences/2013/10/11/fast-scholarship-is-not-always-good-scholarship [Accessed 13 August 2016].

Eckstein, J. (2013) 'Sound reason: Radiolab and the micropolitics of podcasting.' PhD dissertation, University of Denver, Colorado. Available at: http://digitalcommons.du.edu/cgi/viewcontent.cgi?article=1175&context=etd [Accessed 24 May 2016].

Economic and Social Research Council (ESRC) (2016) 'What are the ethical considerations of disseminating findings?' *ESRC.* Available at: www.esrc.ac.uk/funding/guidance-for-applicants/research-ethics/frequently-raised-questions/what-are-the-ethical-considerations-of-disseminating-findings [Accessed 14 November 2016].

Edwards, J. (2015) 'A bank persuaded Twitter to delete my tweets.' *Business Insider.* Available at: http://uk.businessinsider.com/bank-of-america-merrill-lynch-persuaded-twitter-to-delete-my-tweets-2015-12 [Accessed 14 August 2016].

Elbow, P. (2014) 'Maybe academics aren't so stupid after all.' *OUP Blog.* Available at: http://blog.oup.com/2013/02/academic-speech-patterns-linguistics [Accessed 14 August 2016].

Enjolras, B., Steen-Johnsen, K. and Wollebæk, D. (2013) 'Social media and mobilization to offline demonstrations: Transcending participatory divides?' *New Media & Society,* 15(6): 890–908.

Evans, D. (2012) *Social Media Marketing: An Hour a Day.* Chichester: John Wiley & Sons.

Eveleth, R. (2014) 'Academics write papers arguing over how many people read (and cite) their papers.' *Smithsonian,* 25.

Facebook (2016) 'Terms of service.' Available at: www.facebook.com/terms [Accessed 14 August 2016].

Fenton, W. (n.d.) 'Apple iTunes U.' *PC World.* Available at: http://uk.pcmag.com/apple-itunes-u/71352/review/apple-itunes-u [Accessed 3 September 2015].

Fernandes, M., Giesteira, B. and Ai Quintas, A. (2013) 'Visual archives and infographics: New connections.' In N. Ceccarelli (ed.), *2CO COmmunicating COmplexity: 2013 Conference Proceedings.*

Fitzgerald, B. (2015) '*Serial, Mystery Show,* and why listeners want to be in on the investigation.' *Columbia Journalism Review.* October 28. Available at: www.cjr.org/criticism/mystery_show.php [Accessed 24 May 2016].

Fitzpatrick, K. (2015) 'Academia, Not Edu.' *Planned Obsolescence.* October 26. Available at: www.plannedobsolescence.net/academia-not-edu [Accessed 14 November 2016].

FiveThirtyEight (2016) 'Can Carly Fiorina save Ted Cruz's candidacy?' Available at: http://fivethirtyeight.com/features/can-carly-fiorina-save-ted-cruzs-candidacy [Accessed 2 May 2016].

Flukinger, R. (1985) *The Formative Decades: Photography in Great Britain, 1839–1920*. Austin, TX: University of Texas Press.

Fotheringham, M.J., Owies, D., Leslie, E. and Owen, N. (2000) 'Interactive health communication in preventive medicine: Internet-based strategies in teaching and research.' *American Journal of Preventive Medicine*, 19(2): 113–120.

Franklin, T. and Van Harmelen, M. (2007) 'Web 2.0 for content for learning and teaching in higher education.' Report for *JISC*. Available at: www.webarchive.org.uk/wayback/archive/20140614142108/http://www.jisc.ac.uk/media/documents/programmes/digital repositories/web2-content-learning-and-teaching.pdf [Accessed 1 November 2016].

Friesen, N. and Lowe, S. (2012) 'The questionable promise of social media for education: Connective learning and the commercial imperative.' *Journal of Computer Assisted Learning*, 28(3): 183–194.

Fuchs, C. (2013) *Social Media: A Critical Introduction*. London: Sage.

Gainous, J. and Wagner, K.M. (2014) *Tweeting to Power: The Social Media Revolution in American Politics*. Oxford: Oxford University Press.

Gardiner, B., Mansfield, M., Anderson, I., Holder, J., Louter, D. and Ulmanu, M. (2016) 'The dark side of Guardian comments.' *The Guardian*, April 12, sec. Technology. Available at: www.theguardian.com/technology/2016/apr/12/the-dark-side-of-guardian-comments [Accessed 14 November 2016].

Garside, P. (2013) 'Researcher in the field.' [Podcast]. University of Glasgow. Available at: www.gla.ac.uk/researchinstitutes/iii/wtcmp/field [Accessed 5 January 2016].

Garside, P. (2016) Email interview with C. Brumley, 12 January.

Garst, K. (2013) 'Social media as a catalyst for social change.' *Huffington Post*. Available at: www.huffingtonpost.com/kim-garst/social-media-as-a-catalys_b_3197544.html [Accessed 2 May 2016].

Gerbaudo, P. (2012) *Tweets and the Streets: Social Media and Contemporary Activism*. London: Pluto Press.

Gibson, C. (2016) 'Is it possible to air serious academic opinion in 60 seconds or less?' *LSE Communications Blog*. July 26. Available at: http://blogs.lse.ac.uk/communications/2016/07/26/is-it-possible-to-air-serious-academic-opinion-in-60-seconds-or-less [Accessed 14 November 2016].

Gilbert, E., Bergstrom, T. and Karahalios, K. (2009) 'Blogs are echo chambers: Blogs are echo chambers.' In *42nd Hawaii International Conference on System Sciences: HICSS '09*. Waikoloa, HI, USA. pp. 1–10.

Glänzel, W. and Schubert, A. (2005) 'Analysing scientific networks through co-authorship.' In H.F. Moed, W. Glänzel and U. Schmoch (eds), *Handbook of Quantitative Science and*

Technology Research. The Hague: Springer. pp. 257–276. Available at: http://link.springer.com/chapter/10.1007/1-4020-2755-9_12 [Accessed 1 January 2016].

Goldie, S.M. (ed.) (1997) *Florence Nightingale: Letters from the Crimea 1854–1856*. New York: St. Martin's Press.

Goldman, A. and Vogt, P.J. (2015) 'Reply all.' [Podcast.] *Writing on the Wall*. Available at: https://gimletmedia.com/episode/9-yik-yak [Accessed 1 June 2016].

Goodpaster, K.E. (1991) 'Business ethics and stakeholder analysis.' *Business Ethics Quarterly*, 1(1): 53–73.

Google (2016) *Algorithms: Inside Search*. Available at: www.google.co.uk/insidesearch/howsearchworks/algorithms.html [Accessed 14 August 2016].

Gordon, G., Lin, J., Cave, R. and Dandrea, R. (2015) 'The question of data integrity in article-level metrics.' *PLoS Biol*, 13(8): e1002161.

Gordon, L. (2006) 'Dorothea Lange: The photographer as agricultural sociologist.' *The Journal of American History*, 93(3): 698–727.

Grand, A. (2014) 'Peering over the shoulders of giants?' *Science Progress*, 97(4): 383–386.

Greengard, S. (2011) 'Following the crowd.' *Communications of the ACM*, 54(2): 20–22.

Griffel, M. (2012) 'How accurate were Nate Silver's predictions for the 2012 Presidential election? *Forbes*. Available at: www.forbes.com/sites/quora/2012/11/07/how-accurate-were-nate-silvers-predictions-for-the-2012-presidential-election/#56c8d9c928c3 [Accessed 16 August 2016].

Groen, J. and Russo, P. (2015) *The Myth of First-Quarter Residual Seasonality Liberty Street Economics*. Available at: http://libertystreeteconomics.newyorkfed.org/2015/06/the-myth-of-first-quarter-residual-seasonality.html#.VaKKVFpViko [Accessed 2 May 2016].

Grollman, E.A. (2014) 'Blogging for (a) change.' *Conditionally Accepted*. Available at: https://conditionally accepted.com/2014/02/04/blogging-for-a-change [Accessed 14 August 2016].

Grollman, E. (2015) 'For marginalized scholars, self-promotion is community promotion.' Available at: http://blogs.lse.ac.uk/impactofsocialsciences/2015/07/23/self-promotion-imposter-syndrome-marginalized-scholars [Accessed 2 May 2016].

Groth, H. (2012) 'The soundscapes of Henry Mayhew: Urban ethnography and technologies of transcription.' *Cultural Studies Review*, 18(3): 109–130.

Grove, J. (2016) 'Is Snapchat the new Twitter on campus?' *Times Higher Education*. April 21. Available at: www.timeshighereducation.com/news/snapchat-new-twitter-campus [Accessed 14 November 2016].

Grunig, J.E. and Hunt, T.T. (1984) *Managing Public Relations*. New York: Holt, Rinehart & Winston.

Gruzd, A., Staves, K. and Wilk, A. (2011) 'Tenure and promotion in the age of online social media.' *Proceedings of the American Society for Information Science and Technology*, 48(1): 1–9.

Guo, C. and Saxton, G.D. (2013) 'Tweeting social change: How social media are changing nonprofit advocacy.' *Nonprofit and Voluntary Sector Quarterly*, 43(1): 57–79.

Guynn, J. (2015) 'Meet the woman who coined #BlackLivesMatter.' *USA Today*. Available at: www.usatoday.com/story/tech/2015/03/04/alicia-garza-black-lives-matter/24341593 [Accessed 2 May 2016].

Haile, T. (2014) 'What you think you know about the web is wrong.' *Time*. Available at: http://time.com/12933/what-you-think-you-know-about-theweb-is-wrong [Accessed 14 August 2016].

Hall, G. (2015) 'What does Academia_edu's success mean for open access? The data-driven world of search engines and social networking.' *Impact of Social Sciences Blog*. October 22. Available at: http://blogs.lse.ac.uk/impactofsocialsciences/2015/10/22/does-academia-edu-mean-open-access-is-becoming-irrelevant [Accessed 13 August 2016].

Halupka, M. (2014) 'Clicktivism: A systematic heuristic.' *Policy & Internet*, 6(2): 115–132.

Hammersley, B. (2004) 'Audible revolution.' *The Guardian*. February 12. Available at: www.theguardian.com/media/2004/feb/12/broadcasting.digitalmedia [Accessed 24 May 2016].

Hanna, R., Rohm, A. and Crittenden, V.L. (2011) 'We're all connected: The power of the social media ecosystem.' *Business Horizons*, Special issue: *Social Media*, 54(3): 265–273.

Haraway, D.J. (1991) *Simians, Cyborgs, and Women: The Reinvention of Nature*. Abingdon: Routledge.

Harfoush, R. (2009) *Yes We Did! An Inside Look at How Social Media Built the Obama Brand*. Berkeley, CA: New Riders.

Harper, D. (2002) 'Talking about pictures: A case for photo elicitation.' *Visual Studies*, 17(1): 13–26.

Harris, F. (2015) 'The next civil rights movement?' *Dissent*, 62(3): 34–40.

Harrison, B. (2012) *Separate Spheres: The Opposition to Women's Suffrage in Britain* (Vol. 20). London: Routledge.

Hauben, M. and Hauben, R. (1998) 'Behind the net: The untold story of the ARPANET and computer science (Chapter 7).' *First Monday*, 3(8).

Hawn, C. (2009) 'Take two aspirin and tweet me in the morning: How Twitter, Facebook, and other social media are reshaping health care.' *Health Affairs*, 28(2): 361–368.

He, D. and Jeng, W. (2016) 'Scholarly collaboration on the academic social web.' *Synthesis Lectures on Information Concepts, Retrieval, and Services*, 8(1): 1–106.

HEFCE (2016) 'REF impact'. *Higher Education Funding Council for England*. Available at: www.hefce.ac.uk/rsrch/REFimpact [Accessed 13 August 2016].

Hendricks, J.A. and Frye, J.K. (2011) 'Social media and the millennial generation in the 2010 Midterm Election.' In H.S. Noor Al-Deen and J.A. Hendricks (eds), *Social Media: Usage and Impact*. Plymouth, UK: Lexington Books. pp. 183–199.

Himmelstein, D. (2016) 'The history of publishing delays: Satoshi Village.' [Blog post]. Available at: http://blog.dhimmel.com/history-of-delays [Accessed 14 August 2016].

Hofmeyer, A., Newton, M. and Scott, C. (2007) 'Valuing the scholarship of integration and the scholarship of application in the Academy for Health Sciences Scholars: Recommended methods.' *Health Research Policy and Systems*, 5(May): 5.

Holt, K., Shehata, A., Strömbäck, J. and Ljungberg, E. (2013) 'Age and the effects of news media attention and social media use on political interest and participation: Do social media function as leveller?' *European Journal of Communication*, 28(1): 19–34.

House of Commons (2014) HC Debate. July 8, vol. 584, col. 217. Available at: https://hansard.parliament.uk/Commons/2014-07-08/debates/14070874000001/Modern SlaveryBill?highlight=%22jack%20monroe%22#contribution-140708102000081 [Accessed 1 June 2016].

Internet Live Stats (2016) *Total number of websites*. Available at: www.internetlivestats.com/total-number-of-websites [Accessed 27 April 2016].

Ipsos Mori (2016) 'Wellcome Trust Monitor, Wave 3.' Wellcome Trust. April. Available at: http://dx.doi.org/10.6084/m9.figshare.3145744 [Accessed 14 November 2016].

Irani, L. (2015) 'Difference and dependence among digital workers: The case of Amazon Mechanical Turk.' *South Atlantic Quarterly*, 114(1): 225–234.

Jacobson, C. (2001) 'A different way of seeing.' *The Lancet*, 357, May 5: 1454–1455.

Jansen, W. (2009) 'Neurath, Arntz and ISOTYPE: The legacy in art, design and statistics.' *Journal of Design History*, 22(3): 227–242.

Jensen, P., Rouquier, J.B., Kreimer, P. and Croissant, Y. (2008) 'Scientists who engage with society perform better academically.' *Science and Public Policy*, 35(7): 527–541.

Johnson, K. (2007) 'Imagine this: Radio revisited through podcasting.' Unpublished Master's thesis, Texas Christian University, Dallas, TX.

Joseph, S. (2012) 'Social media, political change, and human rights.' *Boston College International & Comparative Law Review*, 35: 145–188.

Kaiser, J. and Fecher, B. (2015) 'Collapsing ivory towers? A hyperlink analysis of the German academic blogosphere.' *LSE Impact of Social Sciences Blog*. September 29. Available at: http://blogs.lse.ac.uk/impactofsocialsciences/2015/09/29/collapsing-ivory-towers-a-hyperlink-analysis [Accessed 1 June 2016].

Kang, C. (2014) 'Podcasts are back – and making money.' *Washington Post*. September 25. Available at: www.washingtonpost.com/business/technology/podcasts-are-back--and-making-money/2014/09/25/54abc628-39c9-11e4-9c9f-ebb47272e40e_story.html [Accessed 24 May 2016].

Kant, V. and Xu, J. (2016) 'News feed FYI: Taking into account live video when ranking feed.' *Facebook Newsroom Blog*. March 1. Available at: https://newsroom.fb.com/news/2016/03/news-feed-fyi-taking-into-account-live-video-when-ranking-feed [Accessed 14 November 2016].

Kaplan, A.M. and Haenlein, M. (2010) 'Users of the world, unite! The challenges and opportunities of social media.' *Business Horizons*, 53(1): 59–68.

Katz, J. S. and Martin, B.R. (1997) 'What is research collaboration?' *Research Policy*, 26(1): 1–18.

Kelleher, K. (2010) 'How Facebook learned from MySpace's mistakes.' *Fortune*. November 19. Available at: http://fortune.com/2010/11/19/how-facebook-learned-from-myspaces-mistakes [Accessed 2 May 2016].

Kendzior, S. (2013) 'The political consequences of academic paywalls.' *Al Jazeera*. Available at: www.aljazeera.com/indepth/opinion/2013/01/2013117111237863121.html [Accessed 14 August 2016].

Kennedy, H. (2015) 'Seeing data: Visualisation design should consider how we respond to statistics emotionally as well as rationally.' *LSE Impact Blog*. July 22. Available at: http://blogs.lse.ac.uk/impactofsocialsciences/2015/07/22/seeing-data-how-people-engage-with-data-visualisations [Accessed 1 June 2016].

Kenski, K., Hardy, B.W. and Jamieson, K.H. (2010) *The Obama Victory: How Media, Money, and Message Shaped the 2008 Election*. Oxford: Oxford University Press.

Khan, S. (2016) '10 ways we are different.' *The Conversation*. Available at: http://theconversation.com/uk/10-ways-we-are-different [Accessed 14 August 2016].

Khondker, H.H. (2011) 'Role of the new media in the Arab Spring.' *Globalizations*, 8(5): 675–679.

Kietzmann, J.H., Hermkens, K., McCarthy, I.P. and Silvestre, B.S. (2011) 'Social media? Get serious! Understanding the functional building blocks of social media.' *Business Horizons*, 54(3): 241–251.

Kimball, M.A. (2006) 'London through rose-coloured graphics: Visual rhetoric and information graphic design in Charles Booth's maps of London poverty.' *Journal of Technical Writing and Communication*, 36(4): 353–381.

Kincaid, D.L. (1985) 'Recent developments in the methods for communication research.' *Journal of East and West Studies*, 14(1): 89–98.

King, G. (2014) 'On Big Data.' *Harvard Magazine*. March. Available at: http://harvardmagazine.com/2014/03/why-big-data-is-a-big-deal [Accessed 1 June 2016].

Kitchin, R., Linehan, D., O'Callaghan, C. and Lawton, P. (2013) 'Public geographies through social media.' *Dialogues in Human Geography*, 3(1): 56–72.

Kjellberg, S. (2010) 'I am a blogging researcher: Motivations for blogging in a scholarly context.' *First Monday*, 15(8).

Knapp, A. (2012) 'ResearchGate wants to be Facebook for scientists.' *Forbes*. March 15. Available at: www.forbes.com/sites/alexknapp/2012/03/15/researchgate-wants-to-be-facebook-for-scientists/#4cb74ecb323f [Accessed 13 August 2016].

Konkiel, S., Sugimoto, C. and Williams, S. (2016) 'The use of altmetrics in promotion and tenure.' *EDUCAUSE Review*, 51(2), March/April. Available at: https://er.educause.edu/articles/2016/3/the-use-of-altmetrics-in-promotion-and-tenure [Accessed 14 November 2016].

Kopf, E.J. (1916) 'Florence Nightingale as statistician.' *Publications of the American Statistical Association*, Vol. 15.

Kramer, B. and Bosman, J. (2015) '101 innovations in scholarly communication – the changing research workflow.' Available at: www.google.com/url?q=https://figshare.com/articles/101_Innovations_in_Scholarly_Communication_the_Changing_Research_Workflow/1286826&sa=D&ust=1471163231820000&usg=AFQjCNFwSbGurNaSs7wlMAf58E-gJ-ZRNg or https://figshare.com/articles/101_Innovations_in_Scholarly_Communication_the_Changing_Research_Workflow/1286826 [Accessed 1 June 2016].

Lain, L.B. (1986) 'Steps toward a comprehensive model of newspaper readership.' *Journalism and Mass Communication Quarterly*, 63(1): 69.

Landman, T. (2012) 'Podcasts.' *Todd Landman*. Available at: www.todd-landman.com/podcasts [Accessed 24 May 2016].

Landman, T. (2016) 'Podcasting is perfect for people with big ideas: Here's how to do it.' *The Guardian*. January 13. Available at: www.theguardian.com/higher-education-network/2016/jan/13/podcasting-is-perfect-for-big-ideas [Accessed 14 November 2016].

Lankow, J., Ritchie, J. and Crooks, R. (2012) *Infographics: The Power of Visual Storytelling*. Hoboken, NJ: Wiley.

Larivière, V., Gingras, Y. and Archambault, É. (2006) 'Canadian collaboration networks: A comparative analysis of the natural sciences, social sciences and the humanities.' *Scientometrics*, 68(3): 519–533.

Latour, B. (2005) *Reassembling the Social: An Introduction to Actor-Network Theory*. Oxford: Oxford University Press.

Lee, S. and Bozeman, B. (2005) 'The impact of research collaboration on scientific productivity.' *Social Studies of Science*, 35(5): 673–702.

Lee, Y.H. and Hsieh, G. (2013) 'Does slacktivism hurt activism? The effects of moral balancing and consistency in online activism.' In *Proceedings of the SIGCHI Conference on Human Factors in Computing Systems*. Paris, France: ACM. April. pp. 811–820.

Leiner, B.M., Cerf, V.G., Clark, D.D., Kahn, R.E., Kleinrock, L., Lynch, D.C., et al. (1997) 'The past and future history of the Internet.' *Communications of the ACM*, 40(2): 102–108.

Lenhart, A. (2012) 'Teens, smartphones & texting.' *Pew Internet & American Life Project*. Available at: www.pewinternet.org/2012/03/19/teens-smartphones-texting [Accessed 14 November 2016]

Lerner, F. (2009) *The Story of Libraries: From the Invention of Writing to the Computer Age*. London: Bloomsbury.

Levitin, D. J. (2014) *The Organized Mind: Thinking Straight in the Age of Information Overload*. London: Penguin.

Lewin, T., Harvey, B. and Page, S. (2012) 'New roles for communication in development?' *IDS Bulletins*. Available at: http://bulletin.ids.ac.uk/idsbo/issue/view/29 [Accessed 9 February 2017].

Lewis, H. (2014) *John Nimmo and Isabella Sorley: A tale of two 'trolls'*. Available at: www. newstatesman.com/media/2014/01/john-nimmo-and-isabella-sorley-tale-two-trolls [Accessed 2 May 2016].

Lewis, H. (2015) 'The echo chamber of social media is luring the Left into cosy delusion and dangerous insularity.' *The New Statesman*. Available at: www.newstatesman.com/ helen-lewis/2015/07/echo-chamber-social-media-luring-left-cosy-delusion-and-dangerous- insularity [Accessed 9 August 2016].

Lillis, T. (1997) 'New voices in academia? The regulative nature of academic writing con- ventions.' *Language and Education*, 11(3): 182–199.

Linton, D. (2014) 'Politics lecturer gives evidence to Parliament.' *News*. Available at: www. manchester.ac.uk/discover/news/article/?id=13538 [Accessed 14 August 2016].

Loehle, C. (1990) 'A guide to increased creativity in research: Inspiration or perspiration?' *BioScience*, 40(2): 123–129.

Long, M.C. (2011) 'Beyond the press release: Social media as a tool for consumer engage- ment.' In H.S. Noor Al-Deen and J.A. Hendricks (eds) *Social Media: Usage and Impact*. Plymouth, UK: Lexington Books. pp.145–159.

Lotan, G., Graeff, E., Ananny, M., Gaffney, D. and Pearce, I. (2011) 'The Arab Spring| the revolutions were tweeted: Information flows during the 2011 Tunisian and Egyptian revolutions.' *International Journal of Communication*, 5: 31.

Lovejoy, K. and Saxton, G.D. (2012) 'Information, community, and action: How non- profit organizations use social media.' *Journal of Computer-Mediated Communication*, 17(3): 337–353.

Lovink, G. (2011) *Networks without a Cause: A Critique of Social Media*. Cambridge: Polity Press.

LSE Impact of Social Sciences Blog (2013) 'Open Access Perspectives in the Humanities and Social Sciences.' Available at: http://blogs.lse.ac.uk/impactofsocialsciences/open- access-ecollection/http://blogs.lse.ac.uk/impactofsocialsciences/open-access-ecollec tion [Accessed 9 February 2017].

LSE Public Policy Group (2011) 'Maximising the impacts of your research: A handbook for social scientists.' *LSE Impact Blog*. Available at: http://blogs.lse.ac.uk/impactofsocialsciences/ the-handbook [Accessed 14 November 2016].

Lupton, D. (2013) 'Why I blog.' *This Sociological Life*. Available at: https://simplysociology. wordpress.com/2013/03/31/why-i-blog [Accessed 14 August 2016].

Lupton, D. (2014) 'Feeling better connected': Academics' use of social media.' Canberra: News & Media Research Centre, University of Canberra. Available at: www.canberra. edu.au/about-uc/faculties/arts-design/attachments2/pdf/n-and-mrc/Feeling-Better- Connected-report-final.pdf [Accessed 1 February 2016].

Lyon, T.P. and Montgomery, A.W. (2013) 'Tweetjacked: The impact of social media on corporate greenwash.' *Journal of Business Ethics*, 118(4): 747–757.

Malik, O. (2006) *Gigaom | Silicon Valley's All Twttr*. Gigaom.com. Available at: https://gigaom.com/2006/07/15/valleys-all-twttr/ [Accessed 2 May 2016].

Malkin, G. (1992) *Who's who in the Internet: Biographies of IAB, IESG and IRSG members*. Available at https://tools.ietf.org/html/rfc1336 [Accessed 14 November 2016].

Mallenbaum, C. (2015) 'The "*Serial* effect" hasn't worn off.' *USA Today*. April 16. Available at: www.usatoday.com/story/life/2015/04/13/serial-podcast-undisclosed/25501075 [Accessed 24 May 2016].

Manjoo, F. (2013) 'You won't finish this article.' *Retrieved*, April 2, p. 2015.

Markman, K.M. (2012) 'Doing radio, making friends, and having fun: Exploring the motivations of independent audio podcasters.' *New Media & Society*, 14: 547–565.

Markman, K.M. and Sawyer, C.E. (2014) 'Why Pod? Further explorations of the motivations for independent podcasting.' *Journal of Radio & Audio Media*, 21(1): 20–35.

Marshall, G.W., Moncrief, W.C., Rudd, J.M. and Lee, N. (2012) 'Revolution in sales: The impact of social media and related technology on the selling environment.' *Journal of Personal Selling & Sales Management*, 32(3): 349–363.

Martinez-Conde, S. (2016) 'Has contemporary academia outgrown the Carl Sagan effect?' *The Journal of Neuroscience*, 36(7): 2077–2082.

Matias, J.N. (2015) 'The tragedy of the digital commons.' *The Atlantic*, June 8. Available at: www.theatlantic.com/technology/archive/2015/06/the-tragedy-of-the-digital-commons/395129 [Accessed 1 June 2016].

Matsakis, L. (2016) 'The unknown, poorly paid labor force powering academic research.' *Motherboard*. Available at: http://motherboard.vice.com/en_uk/read/the-unknown-poorly-paid-labor-force-powering-academic-research [Accessed 13 August 2016].

Maxwell, J.W. (2015) 'Beyond open access to open publication and open scholarship.' *Scholarly and Research Communication*, 6(3): 1–10. Available at: http://src-online.ca/index.php/src/article/view/202 [Accessed 1 June 2016].

Mayhew, H. (1861) *London Labour and the London Poor: A Cyclopaedia of the Condition and Earnings of Those That Will Work, Those That Cannot Work, and Those That Will Not Work* (4 volumes). London: Griffin Bohn.

McCabe, B. (2013) 'Publish or perish: Academic publishing confronts its digital future.' *Johns Hopkins Magazine*, 65(3). Available at: http://hub.jhu.edu/magazine/2013/fall/future-of-academic-publishing [Accessed 1 June 2016].

McCandless, D. (2009) *Information is Beautiful*. Frome: Collins.

McCloskey, A. (2016) 'Using Reddit as a tool for public engagement, profile raising and scholarly dissemination.' *LSE Impact Blog*. Available at: http://blogs.lse.ac.uk/impactofsocialsciences/2016/05/20/using-reddit-as-a-tool-for-public-engagement-profile-raising-and-scholarly-dissemination [Accessed 1 November 2016].

McClung, S. and Johnson, K. (2010) 'Examining the motives of podcast users.' *Journal of Radio & Audio Media*, 17(1): 82–95.

McDonald, S.N. (2015) 'Title of the piece.' *Washington Post*. November 12. Available at: www.washingtonpost.com/news/arts-and-entertainment/wp/2015/11/12/welcome-to-night-vale-creators-chart-the-podcast-path-to-popularity [Accessed 20 February 2016].

McGuigan, G. and Russell, R. (2008) 'The business of academic publishing.' *Electronic Journal of Academic and Special Librarianship*. Available at: http://southernlibrarianship. icaap.org/content/v09n03/mcguigan_g01.html [Accessed 1 June 2016].

McKee, D. and O'Neill, E. (2015) Telephone interview with C. Brumley. November 6.

McKenzie, D.J. and Ozler, B. (2014) 'The impact of economics blogs.' *Economic Development and Cultural Change*, 62(3): 567–597.

McLuhan, M. (1994) *Understanding Media: The Extensions of Man*. Cambridge, MA: MIT Press.

McQuaid, R. (2014) 'Youth unemployment produces multiple scarring effects.' *British Politics and Policy Blog*. Available at: http://blogs.lse.ac.uk/politicsandpolicy/multiple-scarring-effects-of-youth-unemployment [Accessed 14 August 2016].

McQuail, D. (1997) *Audience Analysis*. London: Sage.

McQuillan, D. (2015) 'Bottom-up citizen science projects could challenge authority of orthodox science through community-led investigations.' *LSE Impact of Social Sciences*. January 15. Available at: http://blogs.lse.ac.uk/impactofsocialsciences/2015/01/15/bottom-up-citizen-science-projects-challenge-authority [Accessed 1 June 2016].

Mead, R. (2014) 'The Troll Slayer.' *The New Yorker*. September 1. Available at: www.new yorker.com/magazine/2014/09/01/troll-slayer [Accessed 1 September 2016].

Meadows, R. (2015) 'Dorothea Lange and the art of the caption.' *Contexts: Understanding People in their Social Worlds*, 4(4), Fall.

Menapace, A. (2016) 'Innovative story telling: Data visualisation and interactive platforms.' POLIS Journalism and Crisis Conference. London: UK. April 21.

Mendelow, A.L. (1981) 'Environmental scanning: The impact of the stakeholder concept.' *ICIS 1981 Proceedings*. January. Available at: http://aisel.aisnet.org/icis1981/20 [Accessed 1 November 2016].

Meraz, S. (2009) 'Is there an elite hold? Traditional media to social media agenda setting influence in blog networks.' *Journal of Computer-Mediated Communication*, 14(3): 682–707.

Mewburn, I. and Thomson, P. (2013) 'Why do academics blog? An analysis of audiences, purposes and challenges.' *Studies in Higher Education*, 38(8): 1105–1119.

Meyers, C. (2014) 'Public philosophy and tenure/promotion: Rethinking teaching, scholarship and service.' *Essays in Philosophy*, 15(1).

Mijksenaar, P. (1997) *Visual Function: An Introduction to Information Design*. New York: Princeton Architectural Press.

Miller, D., Costa, E., Haynes, N., McDonald, T., Nicolescu, R., Sinanan, J., et al. (2016) *How the World Changed Social Media*. London: UCL Press.

Mims, C. (2010) 'How Mechanical Turk is broken.' *MIT Technology Review*. Available at: www.technologyreview.com/s/416966/how-mechanical-turk-is-broken [Accessed 13 August 2016].

Mitchell, A. (2013) 'Take back the net: Institutions must develop collective strategies to tackle online abuse aimed at female academics.' *LSE Impact of Social Sciences Blog*. July 24. Available at: http://blogs.lse.ac.uk/impactofsocialsciences/2013/07/24/take-back-the-net-female-academics-online-abuse [Accessed 13 November 2016].

Mobley, E. (1998) 'Ruminations on the sci-tech serials crisis.' *Science and Technology Librarianship*. Available at: www.istl.org/98-fall/article4.html [Accessed 13 August 2016].

Mohdin, A. (2015) 'Academics have found a way to access insanely expensive research papers – for free.' *Quartz*. Available at: http://qz.com/528526/academics-have-found-a-way-to-access-insanely-expensive-research-papers-for-free [Accessed 14 August 2016].

Mollett, A. (2013) 'Using Pinterest to create reading lists: A step by step guide.' *LSE Impact Blog*. September 27. Available at: http://blogs.lse.ac.uk/impactofsocial sciences/2013/09/27/using-pinterest-to-create-reading-lists-a-step-by-step-guide [Accessed 1 June 2016].

Mollett, A. and Fazal, N. (2014) 'Five ways universities are using Instagram.' *LSE Impact Blog*. April 8. Available at: http://blogs.lse.ac.uk/impactofsocialsciences/2014/04/08/five-ways-universities-are-using-instagram [Accessed 1 September 2016].

Mollett, A. and McDonnell, A. (2014) 'Five ways libraries are using Instagram.' *LSE Impact Blog*. April 16. Available at: http://blogs.lse.ac.uk/impactofsocialsciences/2014/04/16/five-ways-libraries-are-using-instagram [Accessed 1 September 2016].

Mollett, A., Moran, D. and Dunleavy, P. (2011) 'Using Twitter in university research, teaching and impact activities. Impact of social sciences: Maximizing the impact of academic research.' *LSE Public Policy Group*. London: London School of Economics and Political Science.

Monroe, J. (2013) 'You can starve on benefits in this country. *The Guardian*. Available at: www. theguardian.com/society/2013/sep/18/jack-monroe-starve-benefits-england [Accessed 2 May 2016].

Moore, S., Gray, J. and Lammerhirt, D. (2016) 'PASTEUR4OA Briefing Paper: Infrastructures for open scholarly communication.' *PASTEUR4OA*. Available at: www.pasteur4oa.eu/sites/pasteur4oa/files/resource/Scholarly%20Platforms%20Briefing%20Paper_FINAL.pdf.bilde [Accessed 14 November 2016].

Moran, M., Seaman, J. and Tinti-Kane, H. (2011) 'Teaching, learning, and sharing: How today higher education faculty use social media.' *Pearson Learning Solutions and Babson Survey Research Group*. Available at: http://files.eric.ed.gov/fulltext/ED535130.pdf [Accessed 13 August 2016]

Morozov, E. (2013) *To Save Everything, Click Here: The Folly of Technological Solutionism.* London: Allen Lane.

Morrison, M. (2016) 'BBC teams up with Viber, WhatsApp to distribute content.' Available at: http://adage.com/article/digital/bbc-teams-viber-whatsapp-distribute-content/302906/ [Accessed 2 May 2016].

Mulligan, C.B. (2013) 'Conflicting pressures on demand for doctors.' *The New York Times.* Available at: http://economix.blogs.nytimes.com/2013/12/11/conflicting-pressures-on-demand-for-doctors/?_r=0 [Accessed 14 August 2016].

Murdock, S. (2015) 'Twitter CEO apologizes for shutting down sites that save politicians' deleted Tweets.' *The Huffington Post.* Available at: www.huffingtonpost.com/entry/twitter-ceo-apol ogizes-politician-tweets_us_5627cff3e4b08589ef4a4b27 [Accessed 14 August 2016].

Murthy, D. (2013) *Twitter: Social Communication in the Twitter Age.* Chichester: John Wiley & Sons.

Myers, G. (1996) 'Strategic vagueness in academic writing.' *Pragmatics and Beyond New Series*, pp. 3–18.

Narisetti, R. (2012) '2011 by the numbers: A memo to Post staff from Managing Editor Raju Narisetti.' *Washington Post.* Available at: www.washingtonpost.com/blogs/ask-the-post/post/2011-by-the-numbers-a-memo-to-post-staff-from-managing-editor-raju-nari setti/2012/01/02/gIQAAGYcWP_blog.html [Accessed 14 August 2016].

NASA (1958) *The National Aeronautics and Space Act of 1958.* Section 203. Washington, DC: Government Printing Office.

NASA (2012) 'Social media at NASA: Slideshare presentation, 2012 edition.' January 7, 2013. Available at: www.slideshare.net/nasa/social-media-at-nasa-2012-edition [Accessed 1 February 2016].

Navarro, C. (2013) 'Hold that thought: Explore a world of ideas.' *Hold That Thought Podcast.* Washington University, St Louis, MO. Available at: http://thought.artsci.wustl. edu [Accessed 14 November 2016].

NCCPE (2013) 'Embedding impact analysis in research.' *JISC.* March. Available at: www.publicengagement.ac.uk/sites/default/files/publication/nccpe_jisc_booklet_ proof_07.05.13.pdf [Accessed 14 November 2016].

Neurath, O. (1931/1991) 'Bildstatistik nach Wiener Methode', *Die Volksschule*, 27: 569 (reprinted in Neurath, O. (1991) *Gesammelte bildpädagogische Schriften*, edited by Rudolf Haller and Robin Kinross. Vienna: Hölder-Pichler-Tempsky, p. 180).

Neurath, O. (1973) 'From Vienna Method to ISOTYPE.' In M. Neurath and R.S. Cohen (eds), *Empiricism and Sociology.* Dordrecht: Reidel.

Newhagen, J.E. and Rafaeli, S. (1996) 'Why communication researchers should study the Internet: A dialogue.' *Journal of Computer-Mediated Communication*, 1(4). Available at: http://onlinelibrary.wiley.com/doi/10.1111/j.1083-6101.1996.tb00172.x/full [Accessed 14 November 2016].

Neylon, C. (2015) 'Disciplinary identities are tightly bound by exclusion: What would scholarship based on inclusion look like?' *Impact of Social Sciences*. January 14. Available at: http://blogs.lse.ac.uk/impactofsocialsciences/2015/01/14/who-do-you-get-to-say-i-am-neylon [Accessed 14 November 2016].

Nightingale, F. (1858) *Notes on Matters Affecting the Health, Efficiency, and Hospital Administration of the British Army. Founded Chiefly on the Experience of the Late War. Presented by Request to the Secretary of State for War.* Privately printed for Miss Nightingale by Harrison and Sons.

NIH (2016) 'A checklist for communicating science and health research to the public.' *National Institutes of Health (NIH)*. May 9. Available at: www.nih.gov/institutes-nih/nih-office-director/office-communications-public-liaison/clear-communication/science-health-public-trust/checklist-communicating-science-health-research-public [Accessed 14 November 2016].

Nyhan, J. and Duke-Williams, O. (2014) 'Is digital humanities a collaborative discipline? Joint-authorship publication patterns clash with defining narrative.' *LSE Impact of Social Sciences*. September 10. Available at: http://blogs.lse.ac.uk/impactofsocialsciences/2014/09/10/joint-authorship-digital-humanities-collaboration [Accessed 14 November 2016].

O'Brien, L. (2006) 'Education goes mobile and multimedia.' *Duke Library Magazine*. Duke University Libraries, 19(2/3). Durham, NC. Available at: http://library.duke.edu/magazine-archive/issue19/feature2.html [Accessed 14 November 2016].

O'Brien, O. and Cheshire, J. (2011) 'The Booth Poverty Map.' [Blog post] *Mapping London*. April 11. Available at: http://mappinglondon.co.uk/2011/the-booth-poverty-map [Accessed 14 November 2016].

O'Connor, K.M., McEwen, L.J. Owen, D. Lynch, K. and Hill, S. (2011) 'Literature review: Embedding public/community engagement in the curriculum – An example of university–public engagement.' Bristol, UK: National Coordinating Centre for Public Engagement. Available at: www.publicengagement.ac.uk/sites/default/files/publication/cbl_literature_review.pdf [Accessed 14 November 2016].

Obar, J.A., Zube, P. and Lampe, C. (2012) 'Advocacy 2.0: An analysis of how advocacy groups in the United States perceive and use social media as tools for facilitating civic engagement and collective action.' *Journal of Information Policy*, 2: 1–25.

Ofcom (2015) *The Communications Market Report*. London: The Office of Communications.

Okerson, A. and J.J. O'Donnell (eds) (1995) *Scholarly Journals at the Crossroads: A Subversive Proposal for Electronic Publishing*. Washington, DC: Office of Scientific & Academic Publishing, Association of Research Libraries.

Oliver, C. (2010) *Parliament and Political Pamphleteering in Fourteenth-Century England*. London: Boydell & Brewer.

Oliver, K. (2014) 'What's the impact of the research impact agenda?' *STEAPP*. May 27. Available at: www.ucl.ac.uk/steapp/steapp-news-publication/2013-14/impact-agenda-oliver [Accessed 13 August 2016].

Ong, W.J. (1975) 'The writer's audience is always a fiction.' *Publications of the Modern Language Association of America*, 90(1): 9–21.

Or, O. (2013) 'Data visualisation and its application in official statistics.' *Proceedings of the 59th ISI World Statistics Congress, 2013*, Hong Kong (Session CPS109), 25–30 August.

Orwell, G. (1946) 'Why I write.' *Gangrel*. Available at: http://orwell.ru/library/essays/wiw/english/e_wiw [Accessed 27 April 2016].

Oser, J., Hooghe, M. and Marien, S. (2013) 'Is online participation distinct from offline participation? A latent class analysis of participation types and their stratification.' *Political Research Quarterly*, 66(1): 91–101.

Ovum.com (2016) *Ovum » Ovum: App Revenue to Double by 2020, Outpacing Download Growth*. Available at: www.ovum.com/press_releases/ovum-app-revenue-to-double-by-2020-outpacing-download-growth [Accessed 2 May 2016].

Owen, L.H. (2016) 'Hoping to make audio more shareable, WNYC introduces "audiograms" for social media'. *NiemanLab*. March 8. Available at: www.niemanlab.org/2016/03/hoping-to-make-audio-more-shareable-wnyc-introduces-audiograms-for-social-media [Accessed 23 April 2016].

Owens, S. (2015) 'Slate's podcast audience has tripled in a year, and its bet on audio over video continues to pay off.' *Nieman Lab*. February 6. Available at: www.niemanlab.org/2015/02/slates-podcast-audience-has-tripled-in-a-year-and-its-bet-on-audio-over-video-continues-to-pay-off/?relatedstory [Accessed 23 April 2016].

Oxford University Press (2016) 'Blog, n.' *Oxford English Dictionary*. [Blog]. Available at: www.oed.com/view/Entry/256732?rskey=rtdVxR&result=1#eid [Accessed 27 April 2016].

Paine, T. (1776) The American Crisis: PHILADELPHIA, January 13, 1777. Available at: www.ushistory.org/paine/crisis/c-02.htm [Accessed 2 May 2016].

Pangburn, E. (2016) *How Many Blogs Are There? | Snitch Blogger – Eric Pangburn*. Available at: http://snitchim.com/how-many-blogs-are-there [Accessed 27 April 2016].

Pasternack, S. and Utt, S. (1990) 'Reader use and understanding of newspaper infographics.' *Newspaper Research Journal*, 11(2): 28–41.

Paton, B. (2013) 'How to use Instagram for research communication.' *Research to Action Blog*. October 22. Available at: www.researchtoaction.org/2013/10/using-instagram-for-research-communication [Accessed 14 November 2016].

Peppler, K. and Solomou, M. (2011) 'Building creativity: Collaborative learning and creativity in social media environments.' *On the Horizon*, 19(1): 13–23.

Pew Research Center (2013) *Social Networking Fact Sheet*. Available at: www.pewinternet.org/fact-sheets/social-networking-fact-sheet [Accessed 2 May 2016].

Pew Research Centre (2015) 'Social media update 2014.' *Pew Research Centre*. Available at: www.pewinternet.org/2015/01/09/social-media-update-2014 [Accessed 1 November 2016].

Phelps, A. (2012) 'Slate doubles down on podcasts, courting niche audiences and happy advertisers.' *Nieman Lab*. June 4. Available at: www.niemanlab.org/2012/06/slate-dou bles-down-on-podcasts-courting-niche-audiences-and-happy-advertisers/?relatedstory [Accessed 14 November 2016].

Pickard, V. and Williams, A.T. (2014) 'Salvation or folly? The promises and perils of digital paywalls.' *Digital Journalism*, 2(2): 195–213.

Playfair, W. (1786) *The Commercial and Political Atlas: Representing, by Means of Stained Copper-Plate Charts, the Progress of the Commerce, Revenues, Expenditure and Debts of England during the Whole of the Eighteenth Century*. London: J Wallis.

Playfair, W. (1801) *Statistical Breviary: Shewing, on a Principle Entirely New, the Resources of Every State and Kingdom in Europe*. London: J Wallis.

Pole, A. (2010) *Blogging the Political: Politics and Participation in a Networked Society*. London: Routledge.

Polonski, V. (2014) 'The evolution of social networking sites: The rise of content-centric platforms which favour the perpetual present.' *LSE Impact of Social Sciences Blog*. January 15. Available at: http://blogs.lse.ac.uk/impactofsocialsciences/2014/01/15/ the-evolution-of-social-network-sites-in-2014 [Accessed 14 November 2016].

Polonski, V. (2015) 'Hacking the system of social influence: How can we use the mechanics of influence to drive behaviour for public good?' *LSE Impact of Social Sciences Blog*. August 7. Available at: http://blogs.lse.ac.uk/impactofsocialsciences/2015/08/07/hacking- the-system-of-social-influence [Accessed 14 November 2016].

Poore, M. (2015) *Using Social Media in the Classroom: A Best Practice Guide*. London: Sage.

Popova, M. (2013) 'Aesthetic consumerism and the violence of photography: What Susan Sontag teaches us about visual culture and the social web.' [Blog post] *Brain Pickings*. September 16. Available at: www.brainpickings.org/2013/09/16/susan-sontag-on-photo graphy-social-media [Accessed 3 February 2016].

Popova, M. (2014) 'Wisdom in the age of information and the importance of story-telling in making sense of the world: An animated essay.' [Blog post] *Brain Pickings*. September 9. Available at: www.brainpickings.org/2014/09/09/wisdom-in-the-age-of-information [Accessed 3 February 2016].

Popper, K.R. (2002) *The Logic of Scientific Discovery*. 2nd Edition. Bristol: Psychology Press (orig. 1959).

Postman, N. (2011) *Technopoly: The Surrender of Culture to Technology*. London: Vintage.

Potter, E. (2010) 'Ducks, decorators, and the dialogical: An examination of approaches to information design.' PhD dissertation, Auckland University of Technology, Auckland, New Zealand.

Preece, J., Maloney-Krichmar, D. and Abras, C. (2003) 'History of emergence of online communities.' *Encyclopedia of Community*. London: Sage. pp. 1–11.

PR Newswire (2015) 'First survey of Serial's listeners sheds light on the Serial Effect'. *PR Newswire*. June 25. Available at: www.prnewswire.com/news-releases/first-survey-of-serials-listeners-sheds-light-on-the-serial-effect-300104734.html [Accessed 30 August 2016].

Putnam, L. (2011) 'The changing role of blogs in science information dissemination.' *Issues in Science and Technology Librarianship*, 65: 4.

Quah, N. (2015a) 'Hot Pod: How Google, the podcast world's big new entrant into the field, will approach curation.' *Hot Pod Newsletter*. Issue 47. *Nieman Lab*. November 3. Available at: www.niemanlab.org/2015/11/hot-pod-how-google-the-podcast-worlds-big-new-entrant-into-the-field-will-approach-curation [Accessed 14 November 2016].

Quah, N. (2015b) 'Hot Pod: Why audio doesn't go viral, revisited.' *Hot Pod Newsletter*. Issue 50. *Nieman Lab*. November 24. Available at: www.niemanlab.org/2015/11/hot-pod-revisiting-the-question-why-doesnt-audio-go-viral [Accessed 14 November 2016].

Quinnell, S. (2011) 'Becoming a networked researcher: Using social media for research and researcher development.' *Impact of Social Sciences Blog*. Available at: http://eprints.lse.ac.uk/51813 [Accessed 14 November 2016].

Quirk, V. (2015) 'Podcast history timeline.' *Knight Lab*. Timeline JS. Available at: http://cdn.knightlab.com/libs/timeline3/latest/embed/index.html?source=1PCTbosxqjGWzq2p84zxqAqAY4rVBjsHJlgFPJUEHR-Q&font=Default&lang=en&initial_zoom=2&height=650 [Accessed 14 November 2016].

Rafaeli, S. (1984) 'The electronic bulletin board: A computer-driven mass medium.' *Social Science Computer Review*, 2(3): 123–136.

Ramirez, R. (1999) 'Stakeholder analysis and conflict management.' In D. Buckles (ed.), *Cultivating Peace: Conflict and Collaboration in Natural Resource Management*. London: International Development Research Centre.

Rapp, A., Beitelspacher, L.S., Grewal, D. and Hughes, D.E. (2013) 'Understanding social media effects across seller, retailer, and consumer interactions.' *Journal of the Academy of Marketing Science*, 41(5): 547–566.

Raynor, D.K., Blenkinsopp, A., Knapp, P., Grime, J., Nicolson, D.J., Pollock, K. et al. (2007) 'A systematic review of quantitative and qualitative research on the role and effectiveness of written information available to patients about individual medicines.' *Health Technology Assessment*, 11(5): 1–160.

Reed, M.S. (2016) *The Research Impact Handbook*. Aberdeen: Fast Track Impact.

Reed, M.S., Graves, A., Dandy, N., Posthumus, H., Hubacek, K., Morris, J. et al. (2009) 'Who's in and why? A typology of stakeholder analysis methods for natural resource management.' *Journal of Environmental Management*, 90(5): 1933–1949.

Reid, R. (2006) 'A visionary and a scoundrel.' [Book review]. *American Scientist*, 94(3), May/June: 274.

Reinert, A. (2011) 'The blue marble shot: Our first complete photograph of Earth. *The Atlantic,* April 12. Available at: www.theatlantic.com/technology/archive/2011/04/the-blue-marble-shot-our-first-complete-photograph-of-earth/237167 [Accessed 14 November 2016].

Remler, D. (2014) 'Are 90% of academic papers really never cited? Reviewing the literature on academic citations.' *Impact of Social Sciences Blog.* April 23. Available at: http://blogs. lse. ac. uk/impact ofsocialsciences/2014/04/23/academic-papers-cita tion-rates-remler [Accessed 2 May 2016].

Resnick, B. (2016) 'Why one woman stole 47 million academic papers – and made them all free to read.' *Vox Science and Health.* Available at: www.vox.com/2016/2/17/11024334/ sci-hub-free-academic-papers [Accessed 14 August 2016].

Reuben, R. (2008) 'The use of social media in higher education for marketing and communi-cations: A guide for professionals in higher education.' Available at: www.fullerton.edu/ technologyservices/_resources/pdfs/social-media-in-higher-education.pdf [Accessed 14 November 2016].

Reuter, C., Heger, O. and Pipek, V. (2013) 'Combining real and virtual volunteers through social media.' *Proceedings of ISCRAM.* Baden-Baden, Germany: Information Systems for Crisis Response and Management. pp. 780–790.

Riemer, K., Richter, A. and Seltsikas, P. (2010) 'Enterprise microblogging: Procrastination or productive use?' *AMCIS 2010 Conference Proceedings.* Lima, Peru: AMCIS. pp. 1–8.

Riismandel, P. (2015) 'Podcast advertisers sharpening focus on listener demographics.' *Midroll.* February 2. Available at: www.midroll.com/podcast-advertisers-focusing-listener-demographics [Accessed 14 November 2016].

Riley, C. (2012) 'Apollo 40 years on: How the moon missions changed the world for ever.' *The Guardian.* December 16. Available at: www.theguardian.com/science/2012/dec/16/ apollo-legacy-moon-space-riley [Accessed 14 November 2016].

Rishika, R., Kumar, A. Janakiraman, R. and Bezawada, R. (2013) 'The effect of customers' social media participation on customer visit frequency and profitability: An empirical investigation.' *Information Systems Research,* 24(1): 108–127.

Roberts, S. (2016) 'Claude Shannon, the father of the information age turns 1100100.' *The New Yorker.* April 30. Available at: www.newyorker.com/tech/elements/claude-shan non-the-father-of-the-information-age-turns-1100100 [Accessed 14 November 2016].

Robinson, T., Patrick, K. and Eng, T. (1998) 'An evidence-based approach to interactive health communication: A challenge to medicine in the information age.' *The Journal of American Medical Association,* 280(14): 1264–1269.

Robinson, W. (2016) '*Serial* season 2 ends this week – exclusive.' *Entertainment Weekly.* March 30. Available at: www.ew.com/article/2016/03/30/serial-season-2-finale-date [Accessed 14 November 2016].

Rogers, E. and Kincaid, L. (1981) *Communication Networks: Toward a New Paradigm for Research.* Berkeley, CA: Free Press.

Rogers, S. (2014) 'What fuels a tweet's engagement?' [Blog post]. *Twitter Blog*. March 10. Available at: https://blog.twitter.com/2014/what-fuels-a-tweets-engagement [Accessed 14 November 2016].

Rohn, J., Curry, S. and Steele, A. (2015) 'UK research funding slumps below 0.5% GDP: Putting us last in the G8.' *The Guardian*, March 13, sec. Science. Available at: www.theguardian.com/science/occams-corner/2015/mar/13/science-vital-uk-spending-research-gdp [Accessed 14 November 2016].

Rosenberg, S. (2010) *Say Everything: How Blogging Began, What it's Becoming, and Why it Matters*. New York, NY: Three Rivers Press.

Ross, P. (2014) 'Photos are still king on Facebook.' [Blog post]. *Social Baker*. April 8. Available at: www.socialbakers.com/blog/2149-photos-are-still-king-on-facebook [Accessed 14 November 2016].

Rothman, J. (2014) 'Why is academic writing so academic?' *The New Yorker*. Available at: www.newyorker.com/books/page-turner/why-is-academic-writing-so-academic [Accessed 14 August 2016].

Rutherford, C. (2010) 'Using online social media to support preservice student engagement.' *Journal of Online Learning and Technology*. Available at: http://jolt.merlot.org/vol6no4/rutherford_1210.htm [Accessed 14 August 2016].

Safko, L. (2009) *The Social Media Bible: Tactics, Tools, and Strategies for Business Success*. Hoboken, NJ: John Wiley & Sons.

Sajuria, J., Hudson, D., Dasandi, N. and Theocharis, Y. (2015) 'Tweeting alone? An analysis of bridging and bonding social capital in online networks.' *American Politics Research*, 43(4): 708–738.

Sasley, B.E. and Sucharov, M. (2014) 'Embracing our (non-scholarly) identities: The benefits of combining engagement with moral activism.' *Impact of Social Sciences Blog*. Available at: http://blogs.lse.ac.uk/impactofsocialsciences/2014/05/01/embracing-our-non-scholarly-identities [Accessed 2 May 2016].

Schradie, J. (2011) 'The digital production gap: The digital divide and Web 2.0 collide.' *Poetics*, 39(2): 145–168.

Selwyn, N. (2012) 'Social media in higher education.' *The Europa World of Learning*. London: Routledge. Available at: www.educationarena.com/pdf/sample/sample-essay-selwyn.pdf [Accessed 14 November 2016].

Settle, J.E., Bond, R.M., Coviello, L., Fariss, C.J., Fowler, J.H. and Jones, J.J. (2015) 'From posting to voting: The effects of political competition on online political engagement.' *Political Science Research and Methods*, 4(2): 361–378.

Shannon, C.E. (1948) 'A mathematical theory of communication.' *Bell System Technical Journal*, 27(3): 379–423.

Shi, F., Foster, J.G. and Evans, J.A. (2015) 'Weaving the fabric of science: Dynamic network models of science's unfolding structure.' *Social Networks*, 43(October): 73–85.

Shirky, C. (2012) 'How we will read.' *Genius.* Available at: http://genius.com/Clay-shirky-how-we-will-read-annotated [Accessed 13 August 2016].

Shortt, N.K., Pearce, J., Mitchell, R. and Smith, K.E. (2016) 'Taking health geography out of the academy: Measuring academic impact.' *Social Science & Medicine,* 168: 265–272.

Sides, J. (2007) 'Why this blog?' *The Monkey Cage.* Available at: http://themonkeycage.org/2007/11/why_this_blog [Accessed 14 August 2016].

Siles-Brügge, G. and De Ville, F. (2013) 'The false promise of EU–US trade talks.' *Manchester Policy Blogs.* Available at: http://blog.policy.manchester.ac.uk/featured/2013/12/the-false-promise-of-eu-us-trade-talks [Accessed 14 August 2016].

Silvertown, J. (2009) 'A new dawn for citizen science.' *Trends in Ecology & Evolution,* 24(9): 467–471.

Simis, M. J., Madden, H., Cacciatore, M.A. and Yeo, S.K. (2016) 'The lure of rationality: Why does the deficit model persist in science communication?' *Public Understanding of Science,* 25(4): 400–414.

Simkin, M.V. and Roychowdhury, V.P. (2002) 'Read before you cite!' *Complex Syst.,* 14: 269–274. Available at: https://arxiv.org/abs/cond-mat/0212043 [Accessed 14 November 2016].

Smiciklas, M. (2012) *The Power of Infographics: Using Pictures to Communicate.* Indianpolis, IN: Que. Available at: http://ptgmedia.pearsoncmg.com/images/9780789749499/samplepages/0789749491.pdf [Accessed 14 November 2016].

Smith, C. (2016) *50+ Amazing Google+ Statistics.* [online]. DMR. Available at: http://expandedramblings.com/index.php/google-plus-statistics [Accessed 2 May 2016].

Smith, S., Ward, V. and House, A. (2011) '"Impact" in the proposals for the UK's Research Excellence Framework: Shifting the boundaries of academic autonomy.' *Research Policy,* 40(10): 1369–1379.

Sontag, S. (1977) *On Photography.* London: Penguin.

Sou, G. (2014) 'Podcasts can "level the playing field" for researchers looking to break the mould and share accessible findings.' *The Impact of Social Sciences Blog.* August 20. Available at: http://blogs.lse.ac.uk/impactofsocialsciences/2014/08/20/viva-voce-podcasts-researchers-level-playing-field [Accessed 14 November 2016].

Southern, L. (2016) 'Analytics nightmare: 82 percent of mobile sharing is done through dark social.' *Digiday.* June 9. Available at: http://digiday.com/publishers/80-percent-mobile-sharing-done-via-dark-social [Accessed 14 November 2016].

Spence, I. (2005) 'No humble pie: The origins and usage of a statistical chart.' *Journal of Educational and Behavioral Statistics,* 30(4), Winter: 353–368.

Spence, I. (2006) 'William Playfair and the psychology of graphs.' *Proceedings of the American Statistical Association, Section on Statistical Graphics.* Alexandria, VA: American Statistical Association. pp. 2426–2436.

Spencer, D.A. (1973) *The Focal Dictionary of Photographic Technologies.* London: Focal Press.

Statista (2016) 'Leading global social networks 2016 | Statistic.' *Statista*. Available at: www.statista.com/statistics/272014/global-social-networks-ranked-by-number-of-users [Accessed 2 May 2016].

Stewart, B. (2016) 'Collapsed publics: Orality, literacy, and vulnerability in academic Twitter.' *Journal of Applied Social Theory*, 1(1). Available at: http://socialtheoryapplied.com/journal/jast/article/view/33 [Accessed 14 November 2016].

Stewart, W. (2000) 'What is ARPANET? The First Internet.' Available at: www.livinginternet.com/i/ii_arpanet.htm [Accessed 2 May 2016].

Stieglitz, S. and Dang-Xuan, L. (2013) 'Social media and political communication: A social media analytics framework.' *Social Network Analysis and Mining*, 3(4): 1277–1291.

Stilgoe, J., Lock, S.J. and Wilsdon, J. (2014) 'Why should we promote public engagement with science?' *Public Understanding of Science*, 23(1): 4–15.

Stokols, D., Hall, K.L. Taylor, B.K. and Moser, R.P. (2008) 'The science of team science.' *American Journal of Preventive Medicine*, 35(2): S77–S89.

Strathern, M. (1997) '"Improving ratings": Audit in the British university system.' *European Review*, 5(3): 305–321.

Street, R.S. (2006) 'Lange's antecedents: The emergence of social documentary photography of California's farmworkers.' *Pacific Historical Review*, 75(3), August: 385–428.

Suber, P. (2012) *Open Access*. Cambridge, MA: MIT Press.

Sugimoto, C. (2016) 'Tenure can withstand Twitter: We need policies that promote science communication and protect those who engage.' *Impact of Social Sciences Blog*. Available at: http://blogs.lse.ac.uk/impactofsocialsciences/2016/04/11/tenure-can-withstand-twitter-thoughts-on-social-media-and-academic-freedom [Accessed 2 May 2016].

Sullivan, A. (2008) 'Why I blog.' *The Atlantic*, 302: 106–113.

Sweney, M. (2016) 'Nick Robinson's *Today* adds nearly 300,000 listeners.' *The Guardian* sec. *Media*. February 4. Available at: www.theguardian.com/media/2016/feb/04/nick-robinson-listeners-bbc-radio-4-today [Accessed 13 August 2016].

Talbot, C. (2014) 'Sir Humphrey and the professors: What does Whitehall want from academics?' Research paper. Manchester: The University of Manchester, Policy @Manchester.

Taylor, M. (2012) 'Academic publishers have become the enemies of science.' *The Guardian*. January 16. Available at: www.guardian.co.uk/science/2012/jan/16/academic-publishers-enemies-science [Accessed 14 August 2016].

Taylor, M. (2013) 'Hiding your research behind a paywall is immoral.' *The Guardian*. January 17. Available at: www.theguardian.com/science/blog/2013/jan/17/open-access-publishing-science-paywall-immoral [Accessed 14 August 2016].

Tech2 (2016) 'World's most popular blogging platform, WordPress turns 10.' *Tech2*. Available at: http://tech.firstpost.com/news-analysis/worlds-most-popular-blogging-platform-wordpress-turns-10-86789.html [Accessed 27 April 2016].

Tench, R. and Yeomans, L. (2009) *Exploring Public Relations*. Second edition. Harlow, UK, and New York: Financial Times/Prentice Hall.

Tervakari, A.M., Silius, K., Tebest, T., Marttila, J., Kailanto, M. and Huhtamäki, J. (2012) 'Peer learning in social media enhanced learning environment.' In *Global Engineering Education Conference (EDUCON)*, 2012 IEEE. April. Washington, DC: Institute of Electrical and Electronics Engineers. pp. 1–9.

Tess, P.A. (2013) 'The role of social media in higher education classes (real and virtual): A literature review.' *Computers in Human Behavior*, 29(5): A60–A68.

Thompson, B. (2013) 'The social/communications map.' *Stratechery*. Available at: https://stratechery.com/2013/socialcommunication-map [Accessed 2 May 2016].

Thomson, J. and Smith, A. (1876) *Street Life in London*. Volume 1. London: Sampson Low, Marston, Searle and Rivington.

Tillery, K. (2012) 'Study of subscription prices for scholarly society journals: Society journal pricing trends and industry overview.' Kansas, MO: Allen Press. Available at: http://allen-press.com/system/files/pdfs/library/2012_AP_JPS.pdf [Accessed November 14 2016].

Tinkler, J. (2015) 'Rather than narrow our definition of impact, we should use metrics to explore richness and diversity of outcomes.' *LSE Impact of Social Sciences Blog*. July 28. Available at: http://blogs.lse.ac.uk/impactofsocialsciences/2015/07/28/impact-metrics-and-the-definition-of-impact-tinkler [Accessed 14 November 2016].

TNS-BMRB and PSI (2015) 'Exploring barrriers to public engagement by UK researchers.' Wellcome Trust. Available at: www.wellcome.ac.uk/About-us/Publications/Reports/Public-engagement/WTP060031.htm [Accessed November 14 2016].

Today (2016) 'Crowdfunding and crowdsourcing in archaeology.' BBC Radio 4. March 4. Available at: www.bbc.co.uk/programmes/b07281zh [Accessed 14 November 2016].

Towne, J. (2012) 'Field gear: Good, better, best.' *Transom.org*. December 23. Available at: http://transom.org/2012/field-gear-good-better-best [Accessed 14 November 2016].

Trench, B. (2008) 'Towards an analytical framework of science communication models.' In D. Cheng, M. Claessens, T. Gascoigne, J. Metcalfe, B. Schiele and S. Shi (eds), *Communicating Science in Social Contexts*. Dordrecht: Springer. pp. 119–135. Available at: http://link.springer.com/chapter/10.1007/978-1-4020-8598-7_7 [Accessed 14 November 2016].

Trottier, D. (2016) *Social Media as Surveillance: Rethinking Visibility in a Converging World*. London: Routledge.

Tsotsis, A. (2016) 'Sean Parker on why MySpace lost to Facebook.' *TechCrunch*. Available at: http://techcrunch.com/2011/06/28/sean-parker-on-why-myspace-lost-to-facebook/ [Accessed 2 May 2016].

Tufte, E.R. (2001) *The Visual Display of Quantitative Information*. Second Edition. Cheshire, CT: Graphics Press.

Tweney, D. (2015) 'Engagement to die for: Snapchat has 100m daily users, 65% of whom upload photos.' *VentureBeat*. Available at: http://venturebeat.com/2015/05/26/snapchat-has-100m-daily-users-65-of-whom-upload-photos/ [Accessed 2 May 2016].

Twitter (2016) 'Terms of service.' Available at: https://twitter.com/tos#yourrights [Accessed 14 August 2016].

UCL (2016a) 'Research images in science communication.' Available at: www.ucl.ac.uk/mathematical-physical-sciences/news-events/science-communication/research-images [Accessed 1 November 2016].

UCL (2016b) 'UCL evaluation framework.' University College London. Available at: www.ucl.ac.uk/public-engagement/evaluation/framework [Accessed 13 August 2016].

University of Manchester, School of Physics and Astronomy (2016) 'The Brian Cox effect.' University of Manchester, School of Physics and Astronomy. Available at: www.physics.manchester.ac.uk/our-research/research-impact/brian-cox-effect [Accessed 30 April 2016].

Urban Dictionary (2016) 'Social media.' *Urban Dictionary*. www.urbandictionary.com/define.php?term=Social+Media&defid=6417897 [Accessed 2 May 2016].

Vaccari, C., Valeriani, A., Barberá, P., Bonneau, R., Jost, J.T., Nagler, J. and Tucker, J.A. (2015) 'Political expression and action on social media: Exploring the relationship between lower-and higher-threshold political activities among Twitter users in Italy.' *Journal of Computer-Mediated Communication*, 20(2): 221–239.

Van der Sar, E. (2015) 'Sci-hub tears down academia's "illegal" copyright paywalls.' [Blog post]. *TorrentFreak*. https://torrentfreak.com/sci-hub-tears-down-academias-illegal-copyright-paywalls-150627/ [Accessed 14 August 2016].

Van Dijck, J. (2013) *The Culture of Connectivity: A Critical History of Social Media*. Oxford: Oxford University Press.

Van Dijck, J. and Poell, T. (2013) 'Understanding social media logic.' *Media and Communication*, 1(1): 2–14.

Van Noorden, R. (2014) 'Online collaboration: Scientists and the social network.' *Nature*, 512(7513): 126–129.

Van Reenen, J. (2016) 'The aftermath of the Brexit vote – the verdict from a derided expert.' *British Politics and Policy at LSE*. August 2. Available at: http://blogs.lse.ac.uk/politicsandpolicy/the-aftermath-of-the-brexit-vote-a-verdict-from-those-of-those-experts-were-not-supposed-to-listen-to [Accessed 14 November 2016].

Van Wingerden, S.A. (1999) *The Women's Suffrage Movement in Britain, 1866–1928*. Basingstoke: Palgrave Macmillan.

Vanwynsberghe, H., Vanderlinde, R. Georges, A. and Verdegem, P. (2015) 'The librarian 2.0: Identifying a typology of librarians' social media literacy.' *Journal of Librarianship and Information Science*, 47(4): 283–293.

Vazquez, L. (2015) 'How NASA turned astronauts into social media superstars.' [Blog post]. *Popular Science*. October 9. Available at: www.popsci.com/how-nasa-trains-astronauts-for-instagram-and-beyond [Accessed 14 November 2016].

Waddington, W. (2015) 'It started with a blog!' *Manchester Policy Blogs*. Available at: http://blog.policy.manchester.ac.uk/posts/2015/09/it-started-with-a-blog [Accessed 14 August 2016].

Wagner, C.S. and Loet, L. (2005) 'Network structure, self-organization, and the growth of international collaboration in science.' *Research Policy*, 34(10): 1608–1618.

Walker, B. (2015) 'Secret histories of podcasting.' *Theory of Everything*. [Podcast]. October 21. Available at: https://toe.prx.org/2015/10/secret-histories-of-podcasting [Accessed 14 November 2016].

Walsh, C. (2011) 'The podcast revolution.' *Harvard Gazette*. October 27. Available at: http://news.harvard.edu/gazette/story/2011/10/the-podcast-revolution [Accessed 14 November 2016].

Walters, K. (2016) '125+ essential social media statistics every marketer should know in 2016.' Available at: https://blog.hootsuite.com/social-media-statistics-for-social-media-managers/ [Accessed 24 August 2016].

Wang, S. (2015) 'Gimlet wants to become the "HBO of podcasting": Here's what its founder's learned trying to get there.' *Nieman Lab*. July 27. Available at: www.niemanlab.org/2015/07/gimlet-wants-to-become-the-hbo-of-podcasting-heres-what-its-founders-learned-trying-to-get-there [Accessed 14 November 2016].

Watters, A. (2013) 'Click here to save education: Evgeny Morozov and ed-tech solutionism.' *Hack Education*. March 26. Available at: http://hackeducation.com/2013/03/26/ed-tech-solutionism-morozov [Accessed 14 November 2016].

Wauters, R. (2010) 'When social media becomes the message: The Gulf oil spill and @BPGlobalPR.' *TechCrunch*. Available at: http://techcrunch.com/2010/06/26/bp-pr-bpglobalpr [Accessed 2 May 2016].

Weiner, J. (2014) 'The voices: Toward a critical theory of podcasting.' *Slate*. December 14. Available at: www.slate.com/articles/arts/ten_years_in_your_ears/2014/12/what_makes_podcasts_so_addictive_and_pleasurable.html [Accessed 14 November 2016].

Wellcome Trust (2016) 'Wellcome Trust Monitor Summary Report Wave 3.' Available at: https://wellcome.ac.uk/sites/default/files/monitor-wave3-summary-wellcome-apr16.pdf [Accessed 14 November 2016].

Weller, M. (2011) *The Digital Scholar: How Technology is Transforming Scholarly Practice*. London: Bloomsbury Academic.

Wen, T. (2015) 'Inside the podcast brain: Why do audio stories captivate?' *The Atlantic*. April 16. Available at: www.theatlantic.com/entertainment/archive/2015/04/podcast-brain-why-do-audio-stories-captivate/389925 [Accessed 14 November 2016].

Whiston Spirn, A. (2008) *Daring to Look: Dorothea Lange's Photographs and Reports from the Field*. Chicago, IL: University of Chicago Press.

Wikipedia (2016) 'Social media.' *Wikipedia*. Available at: https://en.wikipedia.org/wiki/Social_media [Accessed 2 May 2016].

Williamson, A. (2003) 'Audience mapping workbook.' [Working Paper]. *Democratise/ Future Digital*. Available at: http://democrati.se/docs/Democratise.IdentifyingYourStakeholders. pdf [Accessed 14 November 2016].

Wilsdon, J., Allen, L., Belfiore, E., Campell, P., Curry, S., Hill, S., et al. (2015) 'The metric tide: Report of the independent review of the role of metrics in research assessment and management.' *HEFCE*. Available at: www.hefce.ac.uk/pubs/rereports/Year/2015/metric tide/Title,104463,en.html [Accessed 14 November 2016].

Wilson, M. (2015) 'What killed the infographic?' [Blog post]. *FastCoDesign*. June 6. Available at: www.fastcodesign.com/3045291/what-killed-the-infographic [Accessed 14 November 2016].

Wolfsfeld, G., Segev, E. and Sheafer, T. (2013) 'Social media and the Arab Spring politics comes first.' *The International Journal of Press/Politics*, 18(2): 115–137.

Wong, N. (2015) *Talking Back*. Seattle, WA: Red Letter Press.

WordPress.com (2016) 'Notable WordPress users.' *WordPress.com*. Available at: https://wordpress.com/notable-users [Accessed 27 April 2016].

Wren-Lewis, S. (2011) 'Starting a blog, and the role of the ECB in the Euro crisis.' *Mainly Macro*. December. Available at: http://mainlymacro.blogspot.co.uk/2011/12/starting-blog-and-role-of-ecb-in-euro.html [Accessed 14 August 2016].

Wren-Lewis, S. (2015) 'The big picture.' *Mainly Macro*. June 25. Available at: http://mainly-macro.blogspot.co.uk/2015/06/the-big-picture.html [Accessed 14 August 2016].

Wynne, B. (2006) 'Public engagement as a means of restoring public trust in science: Hitting the notes, but missing the music?' *Community Genetics*, 9(3): 211–220.

Xenos, M., Vromen, A. and Loader, B.D. (2014) 'The great equalizer? Patterns of social media use and youth political engagement in three advanced democracies.' *Information, Communication & Society*, 17(2): 151–167.

Xiang, Z. and Gretzel, U. (2010) 'Role of social media in online travel information search.' *Tourism Management*, 31(2): 179–188.

Yang, H. and DeHart, J.L. (2016) 'Social media use and online political participation among college students during the US election 2012.' *Social Media + Society*, 2(1).

Yang, W., Mu, L. and Shen, Y. (2015) 'Effect of climate and seasonality on depressed mood among twitter users.' *Applied Geography*, 63: 184–191.

Yaros, R.A. (2011) 'Social media in education: Effects of personalization and interactivity on engagement and collaboration.' In S. Noor Al-Deen and J.A. Hendricks (eds) *Social Media: Usage and Impact*. Lanham, MD: Lexington Books. pp. 57–174.

Yau Fai Ho, D., Tin Hung Ho, R. and Man Ng, S. (2006) 'Investigative research as a knowledge-generation method: Discovering and uncovering.' *Journal for the Theory of Social Behaviour*, 36(1): 17–38.

Yau, N. (2010) 'The boom of infographics' [Blog post]. *Flowingdata*. May 6. Available at: http://flowingdata.com/2010/05/06/the-boom-of-big-infographics [Accessed 14 November 2016].

Yeo, S. (2013) '"Science is not finished until it's communicated" – UK chief scientist.' *Climate Home – Climate Change News*. October 3. Available at: www.climatechangenews.com/2013/10/03/science-is-not-finished-until-its-communicated-uk-chief-scientist [Accessed 14 November 2016].

Yorke, M. and Yorke, V. (directors) (1977) *The Ho: People of the Rice Pot* [Film].

Zhang, X.S. and Olfman, L. (2010) 'Using blogs to support constructivist and social learning: A case study in a university setting.' [Conference paper] *AMCIS 2010 Proceedings*. pp. 12–15.

Zimmer, C. (2016) 'Staying afloat in the rising tide of science.' *Cell*, 164(6): 1094–1096.

Zorn, E. (2014) 'Smartphones usher in golden age for podcasting.' *Chicago Tribune*. September 20. Available at: www.chicagotribune.com/news/opinion/zorn/ct-smartphones-usher-in-golden-age-podcasts-perspe-20140930-column.html [Accessed 15 February 2016].

INDEX